T0174145

Harvest of Confusion

About the Author

Philip L. Martin earned the B.A. in economics at the University of Wisconsin-Madison in 1971, and the Ph.D. in economics and agricultural economics from the same university in 1975. Since 1975 he has served on the faculty at the University of California-Davis. In 1978-1979, he was a Fellow at the Brookings Institution in Washington, D.C. In 1979-1980, he was senior economist for the Select Commission on Immigration and Refugee Policy and a consultant to the Department of Labor. His publications include: Unfulfilled Promise: Collective Bargaining in California Agriculture (Westview Press, 1988); Migrant Labor in Agriculture: An International Comparison (Giannini Foundation, 1984); and "Labor-Intensive Agriculture" Scientific American October 1983.

Harvest of Confusion: Migrant Workers in U.S. Agriculture

Philip L. Martin
Foreword by John T. Dunlop

Routledge
Taylor & Francis Group

LONDON AND NEW YORK

First published 1988 by Westview Press, Inc.

Published 2018 by Routledge
52 Vanderbilt Avenue, New York, NY 10017
2 Park Square, Milton Park, Abingdon, Oxon OX14 4RN

Routledge is an imprint of the Taylor & Francis Group, an informa business

Library of Congress Cataloging in Publication Data
Martin, Philip L.
  Harvest of confusion.
  (Westview special studies in agriculture science and policy)
  1. Migrant agricultural laborers–United States.
I. Title. II. Series.
HD1525.M23 1988   331.5′44′0973   88-14247
ISBN 13: 978-0-367-01263-2 (hbk)

This book is dedicated to
America's farmworkers,
and the people who
employ, assist, and study them.

Contents

List of Tables and Figures   *ix*
Foreword, John T. Dunlop   *xiii*
Preface   *xv*
Acknowledgments   *xvii*

Chapter 1. Migrant Farmworkers: Background and Definitions   **1**

The Plight of the Migrant Farmworker   4
Migrant Farmworker Definitions   9
Migrant Farmworkers: An Overview   14
Summary   18

Chapter 2. Farm Labor Data   **20**

Concepts: Farms, Farmworkers, and Farm Jobs   21
Establishment Data   26
    Census of Agriculture   26
    Quarterly Agricultural Labor Survey (QALS)   34
    The Farm Costs and Returns Survey   42
    BEA Farm Labor Estimates   45
Household Data   46
    Census of Population   46
    Current Population Survey   49
Administrative Data   54
    ES-202 Employment and Wages Program or UI Data   58
    In-Season Farm Labor Reports (ES-223)   61
Migrant Education   65
Summary   70

Chapter 3.   Bottom-up Migrant Studies   **74**

Migrant Studies Commissioned by the Migrant Health
    Program   77
    The 1973 Migrant Health Program Target Population
        Estimates   78
    The 1978 Migrant Health Target Population Estimates   80
    Methodology for Designating High Impact Migrant and
        Seasonal Agricultural Areas (1985)   84
The Mid-1970s Reports of Lillisand and Rural America   88
Centaur Farmworker Studies for OSHA   94
Summary   98

Chapter 4.   A Top-Down Procedure   **99**

The Number of Migrant Workers   100
The Distribution of Migrants   103
Alternative Approaches   107
Summary   108

Chapter 5.   Migrant Farmworker Characteristics   **110**

Lillisand Migrant Characteristics   112
CPS-HFWF Migrant Characteristics   114
California UI Data   120
Summary   124

Chapter 6.   Migrant Workers in Tomorrow's   **126**
                    Agriculture

The Evolution of Labor-Intensive Agriculture   128
Immigration Reform and Migrant Farmworkers   131
Summary   138

Selected   Bibliography   **140**

Appendix A: Migrant Farmworker Assistance
                    Programs   **155**

Appendix B: Farm Labor Data   **179**

Index   *232*

Tables and Figures

Tables

2.1	Labor Expenditures Reported in the 1982 COA	30
2.2	Hired Worker Employment Reported in the 1982 COA	32
2.3	Fruit, Vegetable, and Hort Specialty Employment and Wages Reported in the 1982 COA	33
2.4	Fieldworker Wages Reported by Farm Labor for July: 1980-1987	41
2.5	Hired Workers on Farms as Reported in Farm Labor: 1985-1987	43
2.6	BEA Estimates of Farm Labor Expenses: 1979-1986	47
2.7	Farmworkers and Farmers in the 1980 Census of Population	50
2.8	Farmworkers and Migrants from the December CPS: 1976-1983	53
2.9	Where CPS-HFWF Migrants Worked: 1981, 1983, 1985	55
2.10	Longest Distance Travelled as a Migrant Farmworker: 1981 and 1983	56
2.11	Farm Employment and Wages Covered by UI in 1985	60
2.12	In-Season Farm Labor Migrant or ES-223 Data: 1978, 1982, 1987	64
2.13	Migrant Education Enrollment and Funding: 1982	69
2.14	Farm Labor Distribution Indicators for the Top Ten States	72
3.1	Migrant Worker and Dependent Factors Used in the 1973 Migrant Health Report	81
3.2	Migrant Health and Lillisand Estimates of Migrant Workers and Dependents for Selected States	82
3.3	Bottom-Up Migrant Studies for 1973 and 1976	95
4.1	Migrant Farmworker Definitions from the 1984 California UI Data	102
4.2	The Distribution of Farmworker Activity: 1982	105
5.1	Migrant Farmworker Characteristics in the December CPS-HFWF 1960-1983	115

Figures

1.1   Defining Migrant Farmworkers   10
2.1   Major Farm Labor Data Sources   25

APPENDIX A: Migrant Farmworker Assistance Programs 155

U.S. Department of Agriculture

Community Facilities Loan Program   156
Farm Labor Housing Loan and Grant Program   157
National School Breakfast Program   158
National School Lunch Program   159
Special Milk Program   160
Special Supplemental Food Program for Women,
  Infants, and Children   161

U.S. Department of Education

College Assistance Migrant Program   162
Handicapped Migratory Agricultural and Seasonal
  Farmworker Vocational and Rehabilitation Services Program   163
High School Equivalency Program   164
Inter/Intrastate Coordination   165
State Basic Grant Program   166

U.S. Environmental Protection Agency

Pesticide Farm Safety Program   167

U.S. Department of Health and Human Services

Community Services Block Grant Program   168
Migrant Head Start   169
Migrant Health Program   170

U.S. Department of Labor

Migrant and Seasonal Agricultural Worker Protection Act
  Program   171
Migrant and Seasonal Farmworker Program (JTPA)   172
Migrant and Seasonal Farmworker Services   173
Temporary Alien Agricultural Labor Certification H-2A
  Program   174

Non-Federal Organizations

Farmworkers Justice Fund   175
Migrant Legal Action Program   176
Interstate Migrant Education Council   177
National Governors Association   178

APPENDIX B: Farm Labor Data   **179**

<u>Tables</u>

A1.1	Farm Sales and Labor Expenditures Reported in the 1982 Census of Agriculture	180
A1.2	Hired Worker Employment Reported in the 1982 Census of Agriculture	182
A1.3	Crop Hired Worker Employment Reported in the 1982 Census of Agriculture	184
A1.4	Vegetable (016) Employment and Wages Reported in the 1982 Census of Agriculture	186
A1.5	Fruit and Nut (017) Employment and Wages Reported in the 1982 Census of Agriculture	188
A1.6	Horticulture Specialty (018) Employment and Wages Reported in the 1982 Census of Agriculture	190
A1.7	Fruit, Vegetable, and Hort Specialty Employment and Wages Reported in the 1982 Census of Agriculture	192
A1.8	Cash Grains (011) Employment and Wages Reported in the 1982 Census of Agriculture	194
A1.9	Other Field (013) Employment and Wages Reported in the 1982 Census of Agriculture	196
A1.10	General Crop (019) Employment and Wages Reported in the 1982 Census of Agriculture	198
A1.11	Average Annual Wages Reported in Farm Labor: 1980, 1985-87	200
A1.12	Average Annual Wages Reported by Farm Labor for Field and Livestock workers: 1980-1987	201
A1.13	Cash Labor Expenses Reported by the Economic Research Service: 1981-84	202
A1.14	Cash Labor Expenses from the FCRS: 1982-86	204
A1.14a	BEA Farm Wages and Labor Expenses for 1982	205
A1.15	BEA Estimates of Farm Labor Expenses: 1979-1984	206
A1.16	Farmers in the 1980 Census of Population	208
A1.17	Farmworkers in the 1980 Census of Population	210
A1.18	Agricultural Employment and Wages Covered by Unemployment Insurance in 1985	212

Tables

A1.19   UI-Covered Employment and Wages Covered in Fruits
        and Vegetables: 1985   216
        UI-Covered Employment and Wages Covered in Fruits
        and Vegetables: 1985(2)   218
        UI-Covered Employment and Wages Covered in Fruits
        and Vegetables: 1985(3)   220
A1.20   ES-223 Data: Migrant Farmworker Estimates for 1982   222
A1.21   ES-223 Data: Seasonal Farmworker Estimates for 1982   224
A1.22   Migrant Education Enrollment and Funding: 1982   226
A1.23   Migrant Estimates Made in 1973 and 1986   228
A1.24   The Distribution of Migrant Worker Activity: 1982 .   230

Foreword

Over the past fifty years there has been an intermittent flow of state and federal government reports on farm labor and on its migratory component. Periodically, attention has centered on housing, pesticides, health care, educational opportunities and on immigration and racial composition. These concerns for deprived workers and their families has been generated on occasion by religious activists, by private foundations, and by the self-help organizing and boycott activities of Cesar Chavez and Baldemar Velasquez. The political process has also generated some of this attention. Very occasionally a serious analytical volume has appeared to interpret agricultural labor markets and to assist in the design of remedial measures, such as Lloyd H. Fisher's The Harvest Labor Market in California (Cambridge, Mass., Harvard University Press, 1953).

Many discussions of agricultural or migratory labor suffer from a gross lack of clarity in defining and measuring farmworkers. This perennial lack of focus and comparability leads to the title of the present study--Harvest of Confusion. The principal contribution of this basic volume is its analysis of definitions and data sources in order to present a reliable overview as well as the state-by-state appendix data.

Martin notes that there are three critical definitions involved in enumerating farmworkers: the definition of a farm, a farmworker, and then a migrant farmworker. Slight changes in definitions lines create very different numbers. Periodic data are obtained from establishments or employers, households or workers, and from the administrative records of government agencies or assistance programs. The matrix of numbers and distribution of migrants thus developed permits useful comparisons among studies and insight into changes over time.

By any indicator, farmworkers are concentrated in a few states; the top ten states include 53 to 87 percent of U.S. migrant activity. California dominates virtually all farmworker rankings, but other states also take on importance depending on definitions and sources. (See the table on p. 72.)

The present volume is intended as a "first building block" to assist in the development of realistic solutions for migrant farmworker issues. The number and distribution of migrants in consistent perspective is an essential requisite to policy discourse.

This volume is an essential beginning to all who are today interested in an overview of agricultural and migratory farm labor and its distribution, and it will prove a helpful tool in bringing together identifiable groups to discuss specialized problems and to design acceptable private and public farm labor policies. It will facilitate forums to discourse, policy proposals, and private and public decisions.

John T. Dunlop
Harvard University
August 1988

Preface

This is a book about migrant farmworkers, some of the most
misunderstood workers in the United States. Countless reports and
exposés have developed the migrant farmworker stereotype familiar
to Americans: migrants are poor minority families who follow the
ripening crops from south to north. As poor and vulnerable
strangers-in-the-fields where they work, they and their children are
underpaid, ill-housed, and subject to abuse by farm employers and
labor contractors. This migrant stereotype is firmly embedded in the
minds of most Americans: a recent Jobs Rated Almanac placed
migrant farmworker last in a list of 250 occupations ranked by
criteria such as salary, job outlook, and work environment.

The truth is that we know remarkably little about the people who
"come with the dust and go with the wind." For example, the
estimated number of migrant farmworkers in the United States
ranges from 100,000 to one million. The profile of a "typical"
migrant is a white college student in one data source, an Hispanic
family in migrant assistance records, and an illegal alien from
Mexico in most worker surveys. There is disagreement over where
migrants live and work: some reports say that migrants live mostly
in California, Florida, and Texas, and then migrate northward in
three streams: other data suggest that these easy-to-picture streams
of migrants moving from south to north are imaginary. Most
Americans believe that migrants are employed only by farmers, but
other data suggest that more and more migrants are employed by
specialized agricultural service firms when they work on farms.

This book fills a lacuna on a subject marked by much debate but
little agreement. It begins with the fundamentals: just who is a
migrant farmworker, and how does the definition of a migrant affect

the number and characteristics of migrant farmworkers? What data sources are available to profile migrant farmworkers, and what are the strengths and weaknesses of each source? What definitions and data did previous studies use to develop the familiar stereotype of poor Hispanic family migrants?

After examining definitions, data, and previous studies, this book estimates the number, characteristics, and distribution of migrants based on the best available data. The book concludes with an assessment of the roles that migrant workers will play in tomorrow's agriculture and how immigration reform may change their roles.

Harvest of Confusion has three purposes. The first is to analyze the vast and diverse data and literature which generate the current confusion over the number and distribution of migrant farmworkers. The second is to suggest a method for counting migrant farmworkers and distributing them across states; the method is suggestive rather than definitive, and is meant to help migrant researchers, administrators, and policymakers make explicit their often implicit assumptions. The third purpose was the tedious but hopeful legacy of this book: the inclusion in the appendices of farm labor data from a variety of sources. This appendix data, in conjunction with the analysis of data sources and previous studies, should be an important starting point for understanding migrant farmworkers.

Philip L. Martin
Davis, California
August 1988

Acknowledgments

This book has evolved from a decade of farm labor research. During this period, my farm labor research program was supported by the Giannini Foundation, the UC Agricultural Issues Center and the UC Davis Agricultural History Center. Some of the data and ideas presented in this book were developed for projects commissioned by the U.S. Department of Labor, the Legal Services Corporation, and California's Employment Development Department. One sponsor deserves special note: the Rosenberg Foundation has been an especially loyal supporter of research on how farmworkers are being affected by immigration reform, and its generosity made possible the presentation of the appendix data.

Stephanie Luce proofread the manuscript and prepared many of the tables in the data appendix, with assistance from Nancy Newsom. Their careful work in compiling and presenting the data is very much appreciated. Joe Bush of the U.S. Department of Labor and Jean Stratford of the UC Davis Institute for Governmental Affairs were very helpful in tracking down elusive references and data. The manuscript was typed under the direction of Lee Knous.

Robert Coltrane, Pat Hogan, Richard Mines, Alan Olmstead, John Staehle, Gary Thompson, and Don Villarejo provided careful reviews of parts of the manuscript. Any errors and omissions are mine and not theirs.

P.L.M.

Chapter 1

Migrant Farmworkers:
Background and Definitions

Migrants are people who move. Migrant farmworkers cross geographic boundaries in order to do farmwork for wages. Most migrants stay away from home for a few weeks or months, forcing them to find temporary accommodations and often disrupting family life. Migrant farmworkers are sometimes vulnerable workers in the areas where they find jobs: many are solo men who travel without their families and whose legal or personal status confines them to farm jobs in remote areas; some are American families for whom migratory farmwork is a last-resort job to survive; and some migrants are workers who could find nonfarm work but prefer to migrate.

Many different individuals are migrant farmworkers, but there are three major groups. Most important are the crews of 20 to 30 immigrant men who are assembled by a crew leader or labor contractor and transported from farm to farm. These young solo men have no families or leave their families behind as they move from farm to farm to harvest crops. As immigrants with little education, an inability to communicate in English, and limited job information and skills which confine them to farmwork, they are vulnerable to unscrupulous labor contractors or crew leaders who (over)charge them for jobs, transportation, and food and housing.

The migrant families memorialized in John Steinbeck's The Grapes of Wrath are the focus of most migrant assistance programs. The stereotypical migrant family packs up its belongings and follows the ripening crops from south to north; along the way, migrant assistance programs provide education and health care for the children and legal services and employment assistance for their parents. The number of migrant families has been declining because there is less temporary family housing available; because many farm employers prefer to hire solo men; and because many migrant families who previously would have migrated have found local farm or non-farm employment.

The third group of migrants is the least studied. Migrant workers also include more skilled and better paid workers who operate grain combines in the midwestern states or apply chemicals to crops in the western states. These workers do not satisfy the migrant stereotype because many are skilled white men, but they satisfy the migrant definition as persons who stay away from home overnight to do farmwork.

The three distinct migrant groups--solo men, migrant families, and skilled migrants--emphasize that migrant farmworker is not an easily-defined occupation filled by workers with similar characteristics and needs. Migrant is really an attribute of **some** of the workers whose occupation is farmworker. The Census of Population, for example, does not include migrant farmworker in its occupational categories, and the Standard Occupational Classification Manual defines general farmworkers (Code 5612), hand vegetable and orchard workers (5613 and 5614), irrigation workers (5615), farm machinery operators (5616), and livestock workers (5617), but not migrant farmworkers. This means that migrant farmworkers are not defined or included in basic occupational data, so that trying to determine the number, distribution, and characteristics of migrant farmworkers is akin to determining how many persons with the occupation salesperson have the "overnight traveller" attribute.

Since migrant farmworker is not a well defined occupation, there are a variety of migrant definitions. Most definitions require a migrant farmworker to cross a geographic boundary and stay away from home overnight to do or look for farmwork. Definitions vary considerably in their treatment of farmworker and farmwork and in determining which boundary must be crossed and what kind of temporary abode must be established to become a migrant farmworker. Workers or families moving from south Texas to Ohio for three months to pick cucumbers are considered migrants under almost all definitions, but what about an Iowa teenager who moves to Kansas for the summer to drive an uncle's tractor for $200 monthly? An alien from Mexico who picks strawberries in California is usually considered a migrant, but what about a Mexican plant pathologist who works on a California farm studying an avocado disease for three months? If the Mexican workers settle in California and then commute daily to their farm jobs, they are no longer migrants under most definitions.

These examples show that many people satisfy the migrant stereotype but not the definition of a migrant (the Mexican family settled in the U.S.), and that other persons satisfy the definition but not the stereotype (the Iowa teenager). The well-entrenched stereotype and the lack of a standard migrant definition makes it hard to enumerate or profile migrant farmworkers. As agriculture evolves, there are fewer south Texas-to-Ohio migrants and more alien workers and skilled migrants, so the characteristics of migrants can change even if their numbers remain the same.

How many migrants are there? No one knows precisely, but journalistic and assistance reports usually refer to "millions" while academic and data-based studies discuss the declining number of migrants. A typical journalistic comment opens Ron Goldfarb's A Caste of Despair:

> "The phenomenon is unique--three streams of people-- eventually hundreds of thousands--flow and fan northward, travelling from their homes around Florida, Texas, and California to distant places, where they will pick and pack crops throughout the agricultural centers of the U.S." (p. 3).

Several pages later, the number is revised upward to "millions of migrant farmworkers hit the road" (p. 9).

Academic studies, on the other hand, emphasize a small and declining number of migrants. Varden Fuller, reflecting on a life of farm labor scholarship in the early 1980s, noted that "the major change that has occurred in respect to seasonal farm labor is the decline in migratoriness (and)...no less important than the decline in physical magnitude is the decline in the myth." (Emerson, p. xi).

Myths die hard, and the myth that there are millions of poor Hispanic migrants lives on in the 1980s. There are indeed poor Hispanic migrants, but there are not millions of them. The data show that many people who satisfy the migrant stereotype (that is, poor minority people) are not in fact migrants, while many of the people who satisfy the definition of a migrant farmworker do not fit the stereotype.

THE PLIGHT OF THE MIGRANT FARMWORKER

Migrant farmworkers have always been a special concern because they were not supposed to be a persisting part of America's family farm system. Congress in the 1862 Homestead Act deemed the family farm system the pillar of rural democracy. During debate on the Homestead Act, there was a call to "keep the plow in the hands of the owner [so that] every new home that is established...is establishing a new republic within the old, and adding a new and strong pillar to the edifice of the state."

Except for plantation agriculture which depended on slaves, America's family farms were expected to supply most of their own labor and capital. The "hired hand" had a place in this family farm agriculture only to the extent that he worked for wages on a neighbor's farm until he moved up the agricultural ladder by buying or homesteading his own farm. Hired hand was only a temporary occupation; since these future farmers were culturally and socially similar (the hired hand could marry the farmer's daughter), farmworkers were believed to be interested primarily in the same high farm prices and low transportation costs that concerned farmers, not higher wages or job security.

A distinct class of migrant farmworkers emerged in the 1850s when midwestern farmers began to specialize in wheat farming instead of planting a mixture of crops adjusted to the availability of family labor. On such large and specialized farms, farmers had to rely on itinerant or migrant workers to harvest the crop. This system of large and specialized farms dependent on hired workers soon reached California, where large and specialized farms developed to take advantage of eastern United States and European markets for wheat and later fruit. Unlike the midwestern stereotype of hired hands being young men employed year-round by older neighbors with a similar culture and outlook, California migrants were travelling strangers who did harvest work for a few weeks and moved on. In California, farms specialized in only one crop, and instead of "'hired men' engaged for a year to work steadily, each one beside a working farmer, now they were extra hands wanted at harvest time only." (Sosnick, p. 4) Since these farms were large, a single farm might employ 2,000 harvest workers, creating a small city on the farm and introducing all the housing, feeding, and law and order issues that accompanied a temporary settlement of workers on a large farmer's property.

Farm employers soon developed the argument that only strangers-in-the-fields could be migrant farmworkers. California farmers argued that they could never hope to employ the normal working population "because California agriculture offered employment only 3 or 4 months in the year--a condition of things entirely unsuited to the demands of the European laborer." This argument soon evolved into familiar cultural stereotypes, such as the argument that "crouching and bending (harvest) operations must be performed in climatic conditions in which only the Orientals and the Mexicans are adapted."

California farmers were very successful in obtaining waves of immigrants who were willing to be migrant farmworkers[1]. The Chinese of the 1880s were followed by the Japanese in the early 1900s, Mexicans and Filipinos after World War I, and Mexicans again during and after World War II. These immigrants were not the only migrant farmworkers; Americans and European immigrants also were migratory farmworkers, but most of them or their children soon managed to escape from farmwork. However, the plight of the migrant was most fully documented during the 1930s, when the Depression forced many Americans to become migrant farmworkers.

The U.S. Senate Committee on Education and Labor (the La Follette Committee) assembled a wealth of materials on migrant farmworkers in California from the 1880s to the 1930s. The Committee concluded that "the economic and social plight of California's agricultural labor is miserable beyond belief" because of low annual earnings, poor housing, and lack of job security (U.S. Senate, 1942, p. 38). Instead of "the deplorable condition of the agricultural labor market [driving] people away from it...the very disorganization of this labor market serves as an attraction to the destitute, the desperate, and the penniless, [so that] California agriculture [serves] as a magnet for the uprooted and the unemployed" (Ibid., p. 39).

The La Follette Committee described in great detail the evolution of large fruit and vegetable farms in California, their "need" for a "roving casual-labor supply" to move from harvest to harvest, and the inability of farmworkers to form trade unions to improve their wages and working conditions. Much of the committee's criticism was directed against the organizations created by farmers, food processors, and bankers "to preserve in themselves an unfettered control of labor and the labor market" (Ibid, p. 39). These employer-oriented organizations were charged with encouraging an over supply of farmworkers in order to hold down wages and, when workers attempted to organize themselves into unions, these employer groups mounted campaigns to label union organizers "radical Communist agitators," to exclude farmworkers from the 1935 National Labor Relations Act which gave most private sector workers the right to join or form unions under government rules, and to work with local law enforcement agencies to frustrate strikes and picketing activities.

The La Follette Committee concluded that federal legislation was needed to protect the economic and civil liberties of migrant farmworkers. The "autocratic system of labor relations [in California agriculture]...provides a clear demonstration that...laissez-faire constitutes too great a danger to the public welfare, too great a drain on the public purse, and too great an injury to the entire fabric of democratic rights" (Ibid, p. 59).

The La Follette Committee report was issued in 1942, when World War II commanded public attention and offered military service and industrial job options to migrant farmworkers. As Americans left the farm labor market, farmers complained of labor shortages, and they argued that the United States would not have enough food to fight World War II unless immigrant farmworkers were imported. Unions and churches protested that there was no shortage of workers, only a shortage of decent wages and working conditions, but farmers obtained permission to import Mexican braceros in 1942, and the bracero program continued to bring several hundred thousand "supplemental" Mexican workers into the farm labor market for the next two decades. Despite the energetic protests of American unions, churches, and activists such as Ernesto Galarza, the bracero program expanded during the 1950s, holding down farm wages and encouraging American workers to find nonfarm jobs.

The 1960s marked a turning point in attitudes toward migrant farmworkers. The 1960's platforms of both the Democratic and Republican parties included statements on migrants: The Democrats pledged "to assure migrant labor, perhaps the most underpriviledged of all, of a comprehensive program to bring them not only decent wages but also an adequate standard of health, housing, social security protection, education, and welfare services." The

Republicans pledged action along "these constructive lines: improvement of job opportunities and working conditions for migratory farmworkers" (U.S. Senate, 1961, p. x). Both houses of Congress created Subcommittees on Migratory Labor which were charged with studying and proposing legislation to help "the forgotten man in modern America; [the] agricultural migrant [who] occupies the lowest rung on the economic and social ladder" (Ibid, p.1).

The plight of the migrant became a national concern during the 1960's. On Thanksgiving day in 1960, CBS television aired *Harvest of Shame*, the best-known exposè of migrant farmworker conditions. In the mid-1960s the bracero program came under sustained attack and was ended, Cesar Chavez began his campaign to organize California farmworkers into the United Farm Workers union, and the federal government initiated programs to provide educational and health service for migrant farmworkers and their children[2]. In the mid-1960s, agricultural economists predicted that a wave of mechanization would eliminate thousands of farm jobs, so "migrant farmworker" became an occupation which required federal assistance to escape. For example, a major review of fruit and vegetable mechanization by economists and engineers concluded in 1970 that mechanization would perpetuate the 1960s "overall surplus of labor" and that the future fruit and vegetable workforce" will consist of a core of skilled and semi-skilled men operating farm machinery supplemented seasonally be unskilled manual laborers, mostly local housewives and school-age youths with a small component of interstate migrant laborers." (Cargill and Rossmiller, p. 18).

The 1960s discussions of migrant farmworkers which led to federal assistance programs simply asserted that there were "millions" of migrants, mostly Black families from Florida and Hispanic families from Texas and California, who packed their children and possessions in cars and drove from south to north. Migrant assistance programs had limited funds, so it did not seem necessary to conduct studies to determine who migrants were or where they moved. To some extent, the late 1960s and early 1970s represented an anomaly in the farm labor market: the termination of the bracero program in 1964 increased the number of American migrant workers to a peak in 1965, but the alternative of nonfarm jobs and welfare pulled and illegal aliens pushed many U.S. workers out of the migratory stream; the 1965 migrant workforce, as measured by the Current Population Survey, was 40 percent larger than the 1967 migrant workforce.

Federal assistance programs began to commission studies of migrants in the 1970s to preserve their funding and to allocate their assistance monies across states. These studies were long on assertion and short on scientific method, so that most of them simply re-affirmed the numbers and especially the allocations of migrants across states which had evolved in the early 1970s. The migrant assistance agencies became interested in the number and distribution of migrants, but each assistance program adopted a different migrant definition and none of these definitions was congruent with the migrant definitions used by non-assistance agencies to collect statistical or administrative data. Thus, counts of migrant farmworkers evolved along two very distinct paths: the Bureau of the Census and the Department of Labor estimated farmworker employment for statistical or administrative purposes, but their definitions and methods were deemed insufficient or unreliable by migrant assistance programs, so each assistance program felt compelled to devise its own counting system or to "adjust" the statistical or administrative data available.

The plight of the migrant became a lower-profile public issue in the 1980s. Migrant assistance agencies developed administrative superstructures that often viewed counting and distributing migrants as a necessary evil in an era when there was little chance to win increases in migrant assistance funds. The quality of statistical and administrative data declined as illegal aliens began became a more significant part of the migrant workforce and as data-gathering resources were curtailed. Researchers began to puncture some of the migrant stereotypes, especially the myth that "millions" of Hispanic families travel with their children from south-to-north, but the migrant family stereotype lives on.

Migrant farmworkers are a diverse group of individuals who, on average, have low annual earnings. However, the diversity of migrants means that many of the past generalizations about the plight of the migrant are no longer appropriate. Some migrants are worse-off in the 1980s than before, such as the non-English and non-Spanish speaking illegal alien migrants from southern Mexico and Guatemala living in rural San Diego county under makeshift tents next to million dollar homes; these migrants earn only the minimum wage when they find work, have no sanitary facilities, and are forced by their illegal status to buy food from roving vendors at often exorbitant prices.

At the other extreme are the semi-skilled and professional migrants. Some semi-skilled migrants in California are employed by large vegetable companies for 8 to 10 months annually and, by migrating between fields owned by their employer and cutting lettuce for piecerate wages of up to $10 hourly, they can earn $15,000 annually. Another group--professional migrants--includes a variety of equipment operators and consultants who sell their specialized services to a number of farmers; some of these professionals harvest crops, while others provide advice on farm pest or financial problems.

The two extremes of solo immigrant men and skilled migrants have been increasing in significance, while the family migrants who were the raison d'etre for migrant assistance programs have been decreasing. In many respects, a declining number of family migrants is a tribute to the success of assistance programs which gave migrants and their children the option of nonfarm jobs. There are still family migrants, but their number and significance appears to have declined.

MIGRANT FARMWORKER DEFINITIONS

Farmworkers work on farms, so the first critical definition is what constitutes a farm, that is, where must persons work or who must employ them for the workers to be farmworkers. Migrants are a subset of farmworkers; once farmworker is defined, it must be determined which geographic boundary a migrant must cross, and how long a migrant farmworker must do farmwork away from home. Figure 1.1 summarizes some of these definition issues.

Three distinct groups of persons work on farms. Farm operators are the persons who own or lease land and put their capital at risk in order to produce farm commodities. They are sometimes assisted by the second group, unpaid family workers such as spouses and children. Farm operators and unpaid family workers work on farms for a share of the farm's profits, not for wages, and it is this method-of-payment which distinguishes them from hired farmworkers. Farm operators and unpaid family workers may migrate to a neighbor's farm and stay away from home overnight to do farmwork on an exchange basis, but if they are not paid wages they are not considered farmworkers and thus cannot be migrants.

Hired workers or farmworkers are persons employed on farms or by farm employers for wages. In most instances, hired workers

Figure 1.1

Defining Migrant Farmworkers

	Term	Definition	Source	Example
1.	Farm	Place which sells farm products worth $1,000 or more	Census of Agriculture	2.2 million farms
2.	Farmworker	Person who does farmwork for wages	Census of Agriculture; QALS	4.9 million "hires" in 1982 COA Peak 1.3 million hired workers employed in July 1987; average 1 million in 1987
		All persons employed on farms	ES-202	44,000 crop and livestock employers in 1986; 616,000 average annual employment
3.	Migrant Farmworker	Crosses county lines and stays away from home overnight to do farmwork for wages	CPS-HFWF	159,000 migrants employed sometime during 1985
		Does 25 to 150 days of farmwork annually, obtains at least half of annual income from farmwork, and cannot return home at the end of a workday	ES-223	765,800 migrants man-months; 1/3 intrastate in 115 reporting areas

are employed at a pre-determined hourly or piecerate wage or for a weekly or monthly salary, although there are continual disputes over whether "sharecroppers" who harvest crops for a share of the crops are hired workers whose minor children cannot work alongside them because child labor laws prohibit their employment or whether sharecroppers are farmers who can bring their children with them into the fields. Agriculture is a large and diverse employer--farmers alone are almost one-seventh of the seven million U.S. employers-- so there is a wide variety of employer-employee relationships.

It is easy to overlook the fact that farm operators and their families do most of the farmwork in the U.S. There are no incontrovertible data, but most studies show that farmers and their families are about 2/3 of all the people employed on farms during a typical week. There are important differences in the roles played by farmworkers: hired farmworkers tend to be most important on large farms that produce fruits and vegetables in California and Florida, while farm operator and unpaid family labor is most important on small and medium-sized livestock and grain farms in the midwest, northeast, and southeast.

Hired workers do a variety of tasks on farms for wages: they harvest crops, they operate equipment, and they sell the farm's products and maintain its books. This means that not everyone employed for wages on a farm has a "farmworker" occupation: in California, which requires virtually all farm employers to participate in the unemployment insurance (UI) system, about one-third of the unemployed workers claiming UI benefits from agricultural employers have nonfarm occupations such as clerk or mechanic.

Hired workers are employed on farms, so a critical definition is what constitutes a "farm". Most migrant farmworker definitions follow the Census of Agriculture (COA): a farm is any place which sold or normally sells $1000 of farm products during the year. Farms can be assigned Standard Industry Codes which reflect the commodity which produces 50 percent or more of the farm's sales, e.g. crop farms (SIC 01) get at least 50 percent of their sales from crops and grape farms (0172) get at least 50 percent of their sales from grapes.

There are three major kinds of "employers" which hire persons to work on farms: crop farms (SIC 01), livestock ranches (02), and agricultural service firms (07).[3] The COA has developed a master list of the 2.2 million U.S. crop and livestock farms, and the COA and sample surveys conducted by USDA with separate lists of farms obtain a variety of employment and wage data from these farms. There is much less data on agricultural service firms, which range from farm labor contractors (SIC 076) to custom lettuce harvesters (072) to dog kennel operators (075) and veterinarians (074) to residential lawn-mowing services (078). Many agricultural service firms employ persons who do farmwork on farms for wages, but these workers are often not reported in the farm-oriented statistical system. For most purposes, the workers employed by agricultural service firms should be eligible to be migrant farmworkers, since a peach picker may be employed directly by the farmer or indirectly through a farm labor contractor.

Agricultural service firms employ a significant but unknown number of "farmworkers", and perhaps half of all migrant farmworkers. California, with its universal UI coverage, provides an illustration: in 1985, about 428,000 workers had at least one job with a firm providing "farm" agricultural services, that is, all agricultural services except pet services (0742 and 0752) and lawn and landscape services (078), versus about 556,000 who had at least one job on a crop farm. Many of the workers employed by these agricultural service firms could be considered migrants: in 1985, about 20 percent of the workers in farm agricultural services had at least two farm employers in two counties, versus only 17 percent of the 556,000 workers employed on California crop farms (some individuals are employed both on crop farms and by agricultural service firms). Thus, the inclusion or exclusion of workers employed by agricultural service firms can affect the number of migrant farmworkers and perhaps their characteristics and distribution.

Packaging and processing farm commodities is an agricultural service which also employs "migrant-type" workers. Once fruits and vegetables are harvested, they must normally be packed or processed. There are three categories of employers which pack and process farm commodities: crop farms (SIC 01), crop preparation services (072), and food manufacturers such as canners (2033) or freezers (2037) of fruits and vegetables. If a commodity such as oranges is grown and packed on the farm, and if more than half the oranges packed were grown on the farm, then the packing shed workers are considered to be "farmworker" employees of a citrus (0174) or crop farm employer. If less than half of the oranges that are packed in a shed owned by the farm operator are grown on the farm, then the packing shed is considered to be a commercial enterprise and the packing shed workers are nonfarm agricultural service employees whose employer is a packing shed (0723). If the oranges are processed into juice, then the processing workers are nonfarm food processing workers (2037). Some migrant assistance programs make only the first group eligible for assistance, some make the first two groups eligible, and some make all three groups eligible.

After defining farm and farmworker, it must be decided which geographic boundary a migrant must cross, what sort of temporary abode must be established, and what proportion of farm and total earnings must come from migratory farmwork. One approach is that developed by USDA to analyze farmworker data from the December Current Population Survey; the CPS-HFWF definition requires a migrant to sometime during the year stay away from home overnight in a different county while doing farmwork for wages.

This seemingly straightforward definition has several implications. First, one night of migratory behavior makes a worker a migrant for one year, so that, in CPS-HFWF reports, a worker who did 100 days of farmwork around his home and then one day in a different county which required an overnight stay is considered to be a migrant who did 101 days of migratory farmwork. Furthermore, this worker's family is considered to be a migrant family for the year; so an Iowa teenage tractor driver who works in Kansas makes his Iowa family a migrant family for that year. Given this inclusive nature of "migrant", it is not surprising that migrant families can include farmworkers, nonfarm workers, and varying numbers of family members.

An alternative migrant definition is that developed by the migrant assistance programs. Migrant Education requires a migrant to cross school district lines in search of farmwork, while Migrant Health does not have a specific boundary crossing requirement but does require a migrant to establish a temporary abode to do farmwork. Migrant Employment and Training programs go still further: migrants are defined as persons who do farmwork for at least 25 but not more than 149 days, obtain at least half of their annual earnings from farmwork, and are unable to return home at the end of a day of farmwork.

Differences in migrant definitions and ambiguity in many of the terms used to define migrants means that there is no easy or straight forward way to enumerate migrants. Since each definition is unique, and no independent statistical data estimate migrants as defined by the various assistance programs, most migrant studies begin with a definition, note that none of the data sources estimates migrants as defined, and then the studies proceed to adjust statistical and administrative data to estimate the number and characteristics of migrants. The result is a wealth of contradictory studies, and a harvest-of-confusion over the "true" picture of migrants.

MIGRANT FARMWORKERS: AN OVERVIEW

One of agriculture's most distinguishing characteristics is concentration: there are over two million farm businesses, or about one in five U.S. businesses is a farm, but the largest 10 percent of all farms produce about two-thirds of the nation's farm products. Farm employment is similarly concentrated: farmworkers are concentrated by size (on large farms), by commodity (on fruit and vegetable farms), and by geography (in California and Florida). Since migrants are a subset of all farmworkers, migrants are similarly concentrated on large fruit and vegetable farms in California and Florida.

This picture of migrants being concentrated on large fruit and vegetable farms in a handful of states when they do farmwork is incontrovertible: what is debated is the exact number of migrants working and living in each state. Most migrant studies estimate the distribution of migrant workers where they work, under the theory that they need most assistance where they are strangers-in-the-land. There are three approaches to determining the distribution of migrants where they work: first, try to count migrants bottom-up in each state to obtain the number and share of migrants in each state; second, begin with aggregate data which reflects migrant worker activity and then distribute this migrant activity across states in a top-down fashion; or third, use migrant service data to indicate the distribution of migrants. The bottom-up approach can yield an estimate of migrants in small geographic areas such as counties, but it suffers from the lack of reliable local data on migrants. The top down approach begins with much better quality employment and labor expenditure data, but such data are only a proxy for migrant activity. The service or client approach freezes the distribution of migrants; wherever service agencies exist, they get clients and thus generate migrants for the area.

The bottom-up approach is used most often by migrant assistance agencies to estimate their target group of migrants. Bottom-up studies conducted in the 1970s typically reported that there were a peak 125,000 to 250,000 migrant workers, and

400,000 to 1 million dependents in migrant households. Some bottom-up studies do not separate migrants and dependents; in such studies, migrant workers are typically 20 percent and dependents 80 percent of the reported migrant population.

The top ten states had about 70 percent of the U.S. migrant population in the three major mid-1970s bottom-up studies (see Chapter 3), but they disagreed on which states had the most migrants. All three studies agreed that California and Florida were among the states with the most migrants, but two reported that Texas had almost one-fourth of all migrants, while the other estimated that Texas had just 4 percent of the migrant population. A notable feature of these studies is the disagreement between their estimates of the migrant population and its distribution: e.g., one study estimated that California had 82,000 or 13 percent of U.S. migrant workers in 1973; another estimated almost three times more or 245,000 (using the same data) or 16 percent of all migrants in 1976; and the third estimated 156,000 or 24 percent of U.S. migrants in 1976.

The three mid-1970s studies relied largely on ES-223 data, or estimates made by Employment Service staff of the number of migrant workers employed during the second week of the month. ES-223 migrant man-months (one migrant employed during the survey week in one month is a migrant man-month) concentrate 80 to 90 percent of the migrants in the top ten states. ES-223 data indicate that California and Florida appear to employ about half of all migrants, and that Texas is conspicuous by its absence among the top ten states. The absence of Texas reflects the fact that ES-223 data count migrants where they work doing migratory farmwork, not where they live, and so Texas-based migrants are counted in Michigan and Ohio, but not in Texas. States also assign different priorities to making ES-223 migrant worker estimates, and Texas may not have attempted to count migrants at work within the state.

Top-down studies, by contrast, give a more reliable picture of the distribution of migrant worker activity across states, but top-down studies provide only a crude guide to the number of migrants in each state. The Census of Agriculture (COA) is one source of top-down data and the 1982 COA indicates that the ten states with the highest labor expenditures and employment accounted for 50 to 80 percent of U.S. farm labor expenditures and employment. According to the COA, California and Florida accounted for about 40 percent of the labor expenditures made by U.S. crop farmers in 1982; these expenditures include wages paid to and payroll taxes paid for family and non family workers and the wages, benefits, and profits of workers hired through crew leaders and labor contractors.

The COA also reports the number of workers employed on farms. The COA data really measure the volume of hiring; the COA records a worker employed each time a person is put on the payroll, so COA cannot determine how many individuals were employed in agriculture because a worker employed on three farms is reported three times. Thus, states such as Kentucky and North Carolina, which employ relatively large numbers of relatives and non-relatives for the 2-to-4 week tobacco harvest, have much larger shares of crop employment than of crop labor expenditures. California, by contrast, accounts for about one-fourth of all crop employment but, because of its relatively high wages and long seasons, California accounts for almost one-third of COA crop labor expenditures.

The COA generates the most reliable labor expenditure data for top-down migrant studies, but data peculiarities mean that they must be interpreted with care. The labor expenditures of farmers indicate both how many workers they hire and what these workers are paid, so one adjustment is to divide labor expenditures by workers to obtain the average labor expenditure per worker hired; such a calculation generates an indicator of each state's relative wages per worker hired on a crop farm. For example, California crop farmers in the 1982 COA reported expenditures of $1.7 billion for 922,000 workers, or $1,824 per worker, while Kentucky crop farmers reported $98 million for 220,000 workers, or $445 per worker. The higher California expenditures per worker reflect higher wages and the tendency of farm jobs to last longer in California than in Kentucky. The COA data can be adjusted to reflect differences in average wages. For example, California fieldworkers who were paid hourly wages averaged $4.69 in July 1982 but only $3.09 in Kentucky; dividing labor expenditures by average wages generates a better measure of the relative volume of farmworker activity in the two states.

Another source of top-down data is the ES-202 data, which reports employment on and wages paid by farms which must provide unemployment insurance for their employees. Beginning in 1978, federal law requires farmers who pay cash wages of $20,000 or more in any calendar quarter or who employ 10 or more workers during 20 weeks of the year to pay UI taxes on the first $7,000 earned by each worker. This 20-10 rule means that an estimated 40 percent of all farmworkers were covered by UI in the mid-1970s, although the concentration of farm employment on larger farms has probably raised UI coverage to cover 70 to 80 percent of the farm wage bill a decade later.

The ES-202 or UI data indicate that covered farm employers, employment, and wages are concentrated in a handful of states. California's share of UI-covered employment is particularly large because California is the only major agricultural state which began to require virtually all farm employers to provide UI coverage for their employees in 1978; Washington, Florida, and Texas have since enacted more-inclusive farmworker coverage.

The UI data are a census of "covered" farm employment and wages. UI data have not been used in top-down farmworker studies because it was assumed that the 20-10 rule excluded "too many" farmworkers outside California. However, a comparison of the 1982 COA and UI data indicate that crop farmers reported a higher total wage bill to UI authorities ($9.3 billion) than they reported to the COA ($6.3 billion), even though farmers presumably had an incentive to underreport farm wages to UI authorities to avoid UI taxes, and they faced no such disincentive when reporting wages to the COA. The UI data suggests that most crop wages are paid by a relative handful of large farms: in 1982, UI data indicate that 32,000 UI-covered crop farm paid an average $148,000 in wages. COA data, by contrast, report that 431,000 crop farms hired workers directly in 1982 and had an average $12,600 in labor expenditures, suggesting that the COA is missing both farm employers and wages paid. In the COA data, half of all crop farm employers produce cash grains and had average $5,900 in labor expenditures.

There is a third approach to count and distribute migrants: simply use intake data from the clients who seek assistance from migrant programs. This approach holds much appeal, since the agency wanting to know about migrants can collect the data it needs. However, most assistance organizations admit that they serve only a fraction of their target clientele, so distributing migrants according to client data merely freezes the distribution of migrants, and makes it hard for an agency to learn that production shifts are introducing migrants into new areas.

The Migrant Education program is the major migrant assistance program, spending about $265 million in FY 1988 to assist the children of migrant farmworkers, and it relies largely, on the number of Full-Time Equivalent (FTE) children in its records in order to allocate assistance funds to states. Migrant Education is the only assistance program which operates its own data base for distribution purposes; Migrant Education spends more to operate this system (most of about $7 million for inter- and intrastate coordination) than is spent on all of the other statistical programs which obtain farm labor data.

SUMMARY

Migrants are the subset of farmworkers who stay away from home to do farmwork for wages. There are three distinct types of migrants: solo men assembled into crews by labor contractors, migrants who travel with their families, and skilled or professional workers who migrate.

The United States was slow to develop a farm labor policy because hired workers were not supposed to be a persisting part of the American family farm system. Instead, "hired hands" were to be only temporary additions to a neighboring farm's family; a person was a hired farmworker only until he saved enough to buy his own farm. The emergence of large and specialized farms in California in the 1880s belied this agricultural ladder, but the myth persisted for almost a century. In the 1960s, the federal government recognized that some farmworkers were "trapped" in agriculture, and launched assistance programs to help migrant workers to escape from farmwork during an era of mechanization.

Defining migrant farmwork requires definitions of farm, farmworker, and migrant. None of these definitions is straight-forward: e.g. should farm be defined by the type of employer or the place where work is performed? Similarly, should farmworkers be defined by the kind of employer who hires them or the work that they do? Finally, what boundary crossing and temporary residence requirements must a migrant farmworker satisfy? There is no consensus on the answers to these questions. Studies which adjust available farm labor data to estimate the number and distribution of migrants answer these questions differently, so it is no surprise that there is widespread disagreement about migrant farmworkers.

There are three major approaches to counting and distributing migrants: begin with local worker data and adjust them bottom-up, begin with statewide expenditure or other proxy data on farmworkers and adjust them top-down, or use client or service data. Most migrant studies use one of the first two approaches, although the largest migrant assistance agency uses the third approach.

NOTES

1. There are several excellent reports and books on the evolution of migrant workers in California agriculture, including Fuller (1939), Daniels (1981), and Jamieson (1945), and McWilliams (1939).

2. The politics of bracero program are explained in Craig (1971), the rise of the UFW is chronicled in Martin et. al. (1988), and the story of migrant assistance programs has not yet been written.

3. Some migrant definitions also include workers employed by forestry firms (SIC 08) and by fishers, hunters, and trappers (SIC 09).

Chapter 2

Farm Labor Data

Defining and enumerating persons who work in U.S. agriculture is complicated by the nature of agricultural production and the characteristics of farmworkers. Agriculture can be limited to the farming units that produce crops and livestock or it can be defined to include the agricultural service firms which e.g., plant and harvest crops but do not make managerial decisions or control the assets on the farm where the work is being performed.

Farmworkers can be defined by what they do or who employs them. Most farm labor data sources define "farmworker" in terms of the employer, not the occupation of the worker or the nature of the work performed. This means that a clerk who is hired directly by the farmer and works on the farm is a "farmworker," but a clerk doing the farm's accounting work in a nonfarm office for a farm accounting service is a nonfarm worker.

Regardless of whether farmworkers are defined by what they do or who employs them, most farmworkers are employed for less than a full year. The seasonality of farmworker employment, exemptions from various labor laws, and the legal and illegal alien status of many farmworkers make it difficult to compile an accurate profile of farmworkers. In particular, seasonality means that the profile of farmworkers in a snapshot survey done in July differs from a profile of year-round workers. Both the profile and the number of workers changes: the July snapshot may yield 100 workers, while

12 monthly snapshots yield an average employment of 40 workers. However, a total of 200 workers may have been hired to generate 100 workers in July and an average 40 throughout the year.

One truism about farm labor data is that no two sources agree on the number, distribution, or characteristics of farmworkers. This does not mean that one source is "correct" and another is "wrong;" instead, differences between data sources reflect the different slices of agricultural employment that each source measures. Agricultural employment estimates vary with the definitions of agriculture, farm employer, and farmworker; the sampling procedure; the number of employers or workers interviewed; and whether farmworker employment refers to the number of persons employed during a snapshot survey, the average number employed over an entire year, or the total number of individuals who worked sometime during the year.

Farm labor data are collected from establishments or employers; households or workers; and data is generated by agencies which administer government programs such as Unemployment Insurance or programs such as Migrant Education which assist farmworkers. This chapter discusses each type of data and then evaluates available farm labor data.

CONCEPTS: FARMS, FARMWORKERS, AND FARM JOBS[1]

All labor data sources must define the group of interest and then develop an enumeration procedure. To enumerate farmworkers, farm employer and farmworker must be defined. In most farm labor data, a farmworker is a person who works for cash wages on a farm (farm operators and unpaid family workers, by contrast, share in the farm's net income), so the first critical definition is what constitutes a farm. Most data sources at least attempt to define a farm as the term is defined in the Census of Agriculture (COA): a farm is any place from which $1,000 or more of agricultural products were sold or normally would have been sold during the year. These COA-defined farms can be assigned to Standard Industrial Classification (SIC) codes that reflect the farm's primary commodity, and the COA divides the farms that satisfy its definition into crop (SIC 01) and livestock (02) categories. Farms are asked to report their sales by commodity, and these sales rankings are used to classify farms into more detailed three digit SIC codes such as fruits and nuts (SIC 017) and four digit codes such grapes (SIC 0172).

Agricultural service firms are not farms, but they often employ workers who do farmwork for cash wages on farms. Agricultural services (SIC 07) include soil preparation (071), crop services (072), and farm labor and management services (076): the service firms in these SIC codes usually employ workers who work for cash wages on a farm, but these "farmworkers" are sometimes included and sometimes excluded from farm labor data.

Farmers have become more and less reliant on agricultural service firms. Most U.S. food and fiber is produced on about 300,000 large farms, and many of these farms hire accountants, operate packing sheds or gins to handle their products, and buy equipment to fertilize or apply chemicals, so that many accountants, packing shed workers and chemical applicators employed on large farms are considered "farmworkers" in data reported by farm operators. However, other large and many smaller farms contract with an agricultural service firm to provide these services. In the case of accountants, such indirect hiring makes them nonfarm workers. The "farmworker" status of other employees depends on several peculiarities, e.g. workers employed in a peach packing shed that packs mostly the peaches of one grower are usually considered farmworkers, but not the workers employed in a shed that packs for a number of growers. Similarly, the peach pickers employed directly by the farmer to pick peaches are considered farmworkers, while peach pickers employed through a farm labor contractor are sometimes considered farmworkers and sometimes not.

Farm employers can be classified as farms or agricultural service firms, and this choice can lead to data ambiguities when workers are asked in a household survey if they worked for wages on a "farm" during the past year. A worker employed by an agricultural service firm such as a labor contractor or a livestock veterinarian may respond in a household survey that he/she worked on a "farm" during the past year and thus be considered a farmworker. However, establishment data such as the COA do not consider agricultural service firms to be farm employers and thus exclude agricultural service "farmworkers," so household and establishment surveys may be reporting different numbers and characteristics of "farmworkers". A "farm" is defined to exclude agricultural service firms in the Census of Agriculture, but "farm" is not defined in the USDA Hired Farm Working Force (HFWF) household survey.

A similar problem arises when "farmworkers" are defined by where they work and not what they do. In establishment data such as the Census of Agriculture and Unemployment Insurance (UI) reports, all persons employed on a farm are farmworkers, including fieldworkers but also clerical and professional staff, the executives of a corporate farm, and paid family members on a family-operated farm. However, if these "farmworkers" are classified by occupation on the basis of what they do, some will be farmworkers but others will be clerks, accountants, and truck drivers. No regularly published data can determine how many persons classified as farmworkers in establishment data are also farmworkers in occupational data, but UI claimant data in several states suggest that only about two-thirds of the persons employed in the industry agriculture have farmworker occupations.

The people who work on "farms" are usually divided into three groups in USDA data: farm operators, unpaid workers, and hired workers. Farm operators are distinguished by working for a share of the profits or a share of the crop and not for an agreed-upon wage, although the tendency of family farms to incorporate for tax and estate reasons has converted some previously self-employed farm operators and unpaid family workers into hired farmworkers.

Unpaid workers are usually family members related to the operator who indirectly benefit from farm profits but are not paid cash wages. Unpaid workers are defined in the USDA's quarterly Agricultural Labor Survey (QALS), for example, as all persons who worked at least 15 hours during the survey week on a farm and were not paid a wage or salary.

Hired farmworkers are all persons who work for wages or a salary on a farm. In most data sources, the minimum time that must be worked for wages is one hour, and one spell (hour) of farmwork makes a person a farmworker for a particular year, even if the person was primarily a student, homemaker, or nonfarm worker during the year. Thus, all persons who had any paid farm employment during the year, including field and livestock workers, equipment operators, bookkeepers, mechanics, and entomologists, veterinarians, and other professionals are considered to be hired farmworkers. If agricultural service firms are also considered farm employers, then e.g., a secretary in the urban office of a crop protection service can be a hired farmworker.

Farm operators, unpaid workers, and hired farmworkers live in single person or family households. The family households pose especially difficult problems for determining the number of dependents in hired farmworker households. In the USDA Hired Farm Working Force (HFWF) reports based on the Current Population Survey (CPS), and under some assistance program regulations, the presence of one person who did qualifying farmwork for wages makes the entire household a farmworker household. The one farmworker may be the household head, the spouse, or a child, so that "farmworker households" often include both farm and nonfarm workers. The migrant subset of hired farmworkers is even more complex: since one migrant farmworker makes the entire household a migrant household, a teenage student in an urban family with parents employed in nonfarm jobs can satisfy the migrant farmworker definition with a summer job and make the entire family a migrant farmworker household for the year.

The final conceptual complication in farm labor data is the difference between farm jobs and farmworkers. Agriculture offers a fluctuating number of jobs, and some farmworkers move from farm to farm and enter and exit the farm workforce during the year. Several data sources estimate the number of workers employed or the number of jobs offered during a particular time period. Workers employed and jobs offered are not identical, even for a survey week, because of worker turnover and various job durations. If worker turnover is high, two or three workers may be hired during the survey week to fill one job slot, so a survey of worker employment during a particular time period must distinguish between total employment (all names on the payroll) and average employment (the average number of workers employed or jobs offered during a one or two week survey period). Survey week jobs may be of different durations: no survey directly distinguishes between a job or worker employed for one hour on one day of the survey week and a job or worker which involves 40 hours. However, labor expenditure data indirectly indicate the duration of employment.

The major farm labor data sources are summarized in Figure 2.1. These data sources are grouped by their purpose and the source of the data. Establishment and household data are drawn from statistical samples of, respectively, employers and households, and the data collected from these samples is assumed to be representative of all employers or households with a known level of reliability. Administrative data, on the other hand, is obtained to operate a program such as the social security system or to generate data on the persons being served by an assistance program.

Figure 2.1
Major Farm Labor Data Sources

Type and Source	Sample Size	Major Data collected	Strengths	Weaknesses
I. Establishment Data				
1. Census of Agriculture	450,000	Employers; Labor expends;* Employ	U.S., State, and County detail	Infrequent; May miss ers
2. Quarterly Agricultural Labor Surv	13,500**	Hourly wages; Hours; Employment	Timely; Only hourly wages	Variance in hourly wages
3. Farm Costs and Returns Survey	24,000	Employer; Labor expenditures;***	Annual by region(NASS) and	Only cost of labor
II. Household Data				
4. Census of Population	19 % samp	Asked occupation last week in March	Has occupation and industry	March is not peak season
5. Current Population Survey	60,000	Small number of farmworkers interviewed		
III. Administrative Data				
A. From Employers				
6. ES-202 UI Data	Census	Includes 40 to 80 percent of all farmwor	Workers, wages, and employmen	Incomplete coverage
7. Social Security	Census	Only 07 employer data published	Workers and wage data	Not published
B. From Workers				
8. Migrant Health	Clients	Demographic and some economic data	Data on persons identified as mig	Not a statistical sample
9. Migrant Education (MSRTS)	Clients	Demographic and some economic data	Data on persons identified as mig	Not a statistical sample
10. DOL MSFW Data (JTPA 402	Clients	Demographic and some economic data	Data on persons identified as mig	Not a statistical sample

*Labor expenditures include gross wages or salaries, employer-paid payroll taxes, the cost of fringe benefits and, for labor contractor, contractor profits.
**The actual sample of respondents is smaller: in July 1986, it was reported that there were 8,500 usable questionnaires. In October 1987, there were 4,400 responses from employers who employed field and livestock workers.
***ERS adjusts the FCRS data and publishes it in the Economic Indicators of the Farm Sector: State Financial Summary.
    SRS publishes the sample data without adjustment in Farm Production Expenditures.

Administrative data from employers is a census of all covered employers, while administrative data from workers profiles only the clients served by a particular program.

ESTABLISHMENT DATA

Establishment labor data are collected from employers, and these employer-reported data typically include total employment at a point in time, such as all persons on the payroll during a survey week; average employment over a period of time; and labor expenditures or the cost of wages and fringe benefits. Establishments can report the method of pay and average hourly earnings of different kinds of workers employed; the number of workers employed or expected to be employed for more and less than e.g., 150 days; and the kinds and costs of fringe benefits which are provided to hired workers. This fringe benefit information may generate proxy data on worker characteristics e.g., employers providing housing for crews of workers may employ migrant domestic or foreign workers, but employers do not report directly whether they hire single or family workers, migrants or local workers, or legal or illegal aliens.

The three major establishment sources of farm labor data are the quinquennial Census of Agriculture (COA), the Quarterly Agricultural Labor Survey (QALS), and reports based on the Farm Costs and Returns Survey (FCRS). Each of these sources is based on a sample of farm employers. The primary purpose of the COA and FCRS is to collect data on how much farmers spend to produce crops and livestock, so both of these sources ask farmers to report their expenditures for labor. The primary purpose of the QALS is to take several "snapshots" of the farm labor market to determine how many people are employed on farms and the hourly wages they are paid.

Census of Agriculture

The Bureau of the Census conducts a number of economic censuses, such as the Census of Manufacturers and the Census of Construction Industries. The Census of Agriculture (COA) is the Bureau's economic census of U.S. farms and ranches. The Census of Agriculture has been conducted periodically since 1840, and is conducted in years that end in 2 and 7, with the questionnaire actually mailed to farmers the year after the reference year, so that the 1987 COA questionnaire was mailed early in 1988. Census data are published for the United States, each State, and for 3100 counties.

The Census of Agriculture is mailed to a continuously updated list of farms maintained by the Bureau of the Census. The COA attempts to enumerate all "places" that sold or normally would have sold farm products worth at least $1,000 in the reference year; in 1982, the COA enumerated 2.2 million farms. Although undercounting, misreporting, and refusing to report are sometimes problems with the Census of Agriculture, the COA is considered the most comprehensive source of farm-related data, especially for states and counties[2]. COA data are often used to benchmark or adjust sample data collected in non-Census years, much as decennial Census of Population data are used to benchmark the monthly Current Population Survey which generates employment and unemployment data.

The COA collects demographic and employment data on one operator per farm; the operator may be the owner, a tenant, a paid manager, or a sharecropper. The COA collects data on the ages and principal occupation of these farm operators, and the 1982 COA reported that the average age of farm operators was 51, that 95 percent of the farm operators were male, and that 45 percent of all farm operators were not primarily farmers, and, furthermore, about 40 percent of all farm operators did 150 or more days of nonfarm work. The COA does not collect data on unpaid family workers.

Farm operator data are collected from all U.S. farms, but farm employment data are obtained from a 20 percent sample of farms, or about 450,000 in 1982. The COA asks this sample of farmers whether they employed farmworkers; whether the workers were employed for more or less than 150 days on their farm; and their expenditures for hired labor, contract labor, and custom work. In the 1982 COA, about 879,000 farms reported hiring almost 5 million workers; labor expenditures were $8.4 billion. Some of these same farms obtain labor through contractors or crew leaders; about 140,000 farms reported expenditures of $1.1 billion for contract labor. Finally, some farm operators rely on custom machine operators to plow or combine, and their expenditures for such custom services include the machine rental fee and the wages of the operator; about 787,000 farm reported expenditures of $2 billion for such custom services in 1982.

The employment data simply ask farmers how many workers they employed for more and less than 150 days during the COA reference year. In 1982, the COA reported that almost 4 million workers were employed less than 150 days on their farms and about 1 million were employed for 150 days or more. This does not mean that the COA reported 4 million "seasonal" and 1 million "regular" workers in 1982; it means only that there were 5 million "hires," so that e.g. a worker employed for 30 days by three different farmers is reported three times in the COA. COA employment data do not distinguish between family and nonfamily hired workers, so that the "regular" and "seasonal" workers reported in the COA might include a son in partnership with a father and teenagers who are paid to work during the summer on their family's farm.

COA employment data can be cross-tabulated in a variety of ways. For example, the volume of employment can be reported by commodity: in 1982, there were 33,400 fruit, vegetable, and horticultural specialty farms that employed workers for 150 days or more, and they employed 279,000 or 30 percent of all "regular" workers reported. About 64,000 FVH farms employed workers for less than 150 days, and they employed 1.2 million or 30 percent of all "seasonal" workers.

The Census of Agriculture has dropped the workers-employed question from the 1987 COA because USDA has adopted a total cost-of-production approach to farming which does not permit lower-priority hired worker questions to be included on the questionnaire. This means that instead of asking farmers how many workers they employed, USDA is interested primarily in farmers' expenditures for labor. Labor expenditures include gross wages or salaries, employer-paid payroll taxes, and the costs of any fringe benefits such as health insurance offered to workers. Labor expenditure data alone provide only limited information on farmworkers because they include expenditures for family labor, and thus may change as tax laws make it more or less advantageous to incorporate or to pay children for doing farmwork, and because labor expenditures may rise while wages remain stable or even fall if, for example, social security or other payroll taxes rise or if fringe benefit costs rise.

Despite these deficiencies, COA labor expenditure data is the best data available by state and county which can distribute farmworkers across geographic areas. Hours worked are not reported in the COA, , simply dollar costs, but if reported labor expenditures are divided by an independent average hourly wage, approximate hours worked can be estimated. For example, California farmers reported paying 31 percent of the labor expenditures of 1982 COA crop farms, and Florida farmers paid almost 9 percent. However, the average fieldworker wage in California was $4.69 in July 1982, versus $3.90 in Florida. If crop labor expenditures are divided by these hourly wages to standardize them for wage differences, California's share of crop hours worked falls to 26 percent and Florida's rises toward 14 percent.

Labor expenditure data can be cross-tabulated by commodity, state, and size of expenditure. In the 1982 COA, farmers reported labor expenditures of $8.4 billion, and these expenditures were divided roughly into thirds, with grain, field and general crop farms reporting $2.6 billion; fruit, vegetable, and horticultural specialty (FVH) farms $2.8 billion, and livestock farms $3.0 billion. The COA reports how many farms paid e.g. more and less than $50,000, but not what they paid, so it is not possible to determine just how concentrated labor expenditures are. Instead, the COA reports labor expenditures by farm sales, and in 1982, the 6,500 farms which each sold farm products worth $1 million or more and each had labor expenditures of $50,000 or more reported total labor expenditures of about $2.7 billion; meaning that less than 1 percent of all farm employers accounted for one-third of all labor expenditures.

Farmers also report their expenditure for contract labor and for custom work such as combining grain or fertilizing. Expenditures for contract labor represent the wages, payroll taxes, fringe benefits, and profits of workers employed by a labor contractor or crewleader. Expenditures for custom work are distinguished from contract labor expenditures by the machine rental fee: custom work implies that the expenditure is primarily paying for equipment rental, not labor time, although a farmer paying for custom combine services is paying for both the equipment and labor. Between 1978 and 1982, labor expenditures rose 19 percent, contract labor rose 22 percent, and custom work increased 15 percent.

Labor expenditures are concentrated in a few states. Table 2.1 ranks states by labor expenditures reported by farms, and it indicates that the top 10 states reported 54 percent of all labor expenditures. Contract labor expenditures are even more concentrated in a handful

Table 2.1

Labor Expenditures Reported in the 1982 Census of Agriculture

State	Crop&Live Labor Expend($000)	Per Dist	Crop&Live Contract Labor Expend($000)	Per Dist	Crop&Live Total Labor Expend($000)	Per Dist	Crop Labor Expend($000)	Per Dist	Crop Contract Lab Expend($000)	Per Dist	Crop Tot Labor Expend($000)	Per Dist
California	1,819,323	22%	413,766	37%	2,233,089	23%	1,681,436	31%	400,002	42%	2,081,438	33%
Texas	480,462	6%	88,334	8%	568,796	6%	274,717	5%	55,389	6%	330,106	5%
Florida	480,444	6%	201,298	18%	681,742	7%	472,408	9%	196,507	21%	668,915	11%
Washington	313,100	4%	33,501	3%	346,601	4%	262,910	5%	31,897	3%	294,807	5%
Wisconsin	279,154	3%	9,988	1%	289,142	3%	61,758	1%	5,087	1%	66,845	1%
New York	246,022	3%	12,778	1%	258,800	3%	90,809	2%	10,113	1%	100,922	2%
North Caroli	245,364	3%	20,747	2%	266,111	3%	183,623	3%	17,561	2%	201,184	3%
Illinois	225,820	3%	8,024	1%	233,844	2%	134,405	2%	5,876	1%	140,281	2%
Pennsylvani	224,174	3%	12,249	1%	236,423	2%	44,645	1%	9,297	1%	53,942	1%
Iowa	222,146	3%	8,358	1%	230,504	2%	91,027	2%	6,646	1%	97,673	2%
Top Ten Stat	4,536,009	54%	809,043	73%	5,345,052	56%	3,297,738	61%	738,375	78%	4,036,113	64%
U.S. Total	8,441,180	100%	1,103,773	100%	9,544,953	100%	5,408,782	100%	945,062	100%	6,353,844	100%

Source: Census of Agriculture, 1982; a farm is a place which sold farm products worth $1000 or more in 1982

Labor expenditures include gross wages or salaries paid to family and nonfamily hired workers and supervisors, bonuses,

  social security and other payroll taxes, and expenditures for fringe benefits

Contract labor expenditures include wages paid to hired workers and supervisors, payroll taxes and fringe benefits, of states; three-quarters are reported by farms in the top 10 states. California clearly accounts for the most labor expenditures; about one-fourth of COA labor expenditures, and California, Texas, and Florida account for over one-third of all labor expenditures.

Migrants are associated with crop farms, and the labor expenditures of crop farms are even more concentrated in a few states. California alone accounts for one-third of all crop labor expenditures, and California and Florida account for 40 percent. Note that these two states and Texas and Washington account for 54 percent of crop labor expenditures.

Hired workers are even more concentrated in a few states. Table 2.2 ranks the top 10 states by the number of workers hired less than 150 days on the responding farm in 1982, and these states included almost three-quarters of all "seasonal" workers hired. This table is surprising because a state such as Florida reports fewer seasonal workers hired than Kentucky and Tennessee, and other significant "migrant" states such as Michigan and Ohio do not appear among the top 10. There are several reasons for this ranking of states: California has almost three times more seasonal and regular workers than other states, and it includes one-third of the regular workers hired and one-sixth of the seasonal workers hired.

Kentucky and Tennessee appear on the top 10 employment list (but not on the top 10 labor expenditure list) because these states include large numbers of small farm employers which each report hiring a few workers, e.g. Kentucky's 50,000 farm employers are second only to Texas' 63,000, and Tennessee's 36,000 are close to California's 40,000. However, most of these small farm employers report low expenditures and seasonal workers, e.g. Kentucky farm employers report an average $3,600 in total labor expenditures, versus $55,800 in California. The suspicion that much of this Appalachian farm employment represents small farms paying relatives for very short seasonal jobs is bolstered by unemployment insurance data: only 233 Kentucky crop and livestock farmers paid UI taxes in 1985, and they reported only $50 million in covered wages, versus over 700 employers and $120 million wages in a state such as Michigan.

Migrant workers are most closely associated with fruits, vegetables, and horticultural specialties (FVH). As Table 2.3 indicates, FVH labor expenditures and employment are concentrated in a handful of states. California accounts for almost half of total FVH labor expenditures (those made directly to workers and those made through labor contractors), and California and Florida account

Table 2.2

Hired Worker Employment Reported in the 1982 Census of Agriculture

State	Crop&Live Workers	Per Dist	Crop&Live Seasonal Workers	Per Dist	Crop&Live Regular Workers	Per Dist	Total Crop Workers	Per Dist	Crop Seasonal Workers	Per Dist	Live Seasonal Workers	Per Dist
California	979,874	20%	640,750	16%	339,510	36%	921,620	28%	607,601	22%	25,491	7%
Kentucky	287,064	6%	247,808	6%	39,352	4%	219,532	7%	189,009	7%	8,829	2%
North Caroli	271,293	6%	233,420	6%	38,109	4%	237,769	7%	207,800	8%	8,140	2%
Washington	259,202	5%	230,121	6%	29,195	3%	228,461	7%	205,624	8%	6,358	2%
Texas	236,500	5%	173,712	4%	62,976	7%	119,069	4%	84,227	3%	28,134	7%
Iowa	206,130	4%	146,776	4%	55,627	6%	115,455	4%	77,510	3%	17,682	5%
Tennessee	208,727	4%	139,244	4%	70,148	7%	149,811	5%	85,870	3%	6,207	2%
Florida	235,381	5%	129,316	3%	106,106	11%	209,967	6%	112,293	4%	8,432	2%
Minnesota	182,779	4%	126,435	3%	56,417	6%	101,243	3%	62,983	2%	18,157	5%
Oregon	165,254	3%	120,408	3%	44,993	5%	140,616	4%	101,025	4%	5,402	1%
Top Ten Stat	3,032,204	63%	2,187,990	56%	842,433	89%	2,443,543	74%	1,733,942	64%	132,832	34%
U.S. Total	4,839,112	100%	3,906,376	100%	943,943	100%	3,287,263	100%	2,729,388	100%	386,068	100%

Source: Census of Agriculture, 1982; a farm is a place which sold farm products worth $1000 or more in 1982

Seasonal workers were employed less than 150 days on the responding farm; a worker employed on two farms
  is counted twice in this data

Regular workers were employed on the responding farm more than 150 days

Table 2.3

Fruit, Vegetable, and Hort Specialty Employment and Wages Reported in the 1982 COA

State	Total FVH Farm Ers	Per Dist	Total Labor Expend -($000)-	Average Labor Expen -($000)-	Per Dist	FVH Regular Workers	Per Dist	FVH Seasonal Workers	Seasonal-Per of Tot Workers	Per Dist
California	23,713	33%	1,581,689	67	46%	116,910	42%	534,613	82%	44%
Florida	5,880	8%	486,149	83	14%	39,149	14%	96,895	71%	8%
Washington	5,708	8%	179,765	31	5%	12,366	4%	167,478	93%	14%
Pennsylvania	2,085	3%	119,999	58	3%	11,918	4%	21,643	64%	2%
Texas	2,482	3%	98,175	40	3%	8,110	3%	20,647	72%	2%
New York	3,408	5%	95,747	28	3%	8,045	3%	38,940	83%	3%
Michigan	3,835	5%	93,744	24	3%	7,414	3%	55,369	88%	5%
Oregon	2,902	4%	88,376	30	3%	6,186	2%	72,762	92%	6%
Arizona	460	1%	82,333	179	2%	5,968	2%	17,907	75%	1%
New Jersey	1,399	2%	57,564	41	2%	5,891	2%	14,698	71%	1%
Top Ten Stat	51,872	72%	2,883,541		83%	221,957	80%	1,040,952	71%	85%
United States	72,388	100%	3,467,575	48	100%	277,955	100%	1,222,915	81%	100%

Source: Bureau of the Census, Census of Agriculture, 1982; a farm is a place which sold farm products worth $1000 or more in 1982

Labor expenditures include gross wages or salaries paid to hired workers and supervisors, bonuses, social security and other payroll taxes, and expenditures for fringe benefits

Large farm employers reported paying $50,000 or more in 1982; large contract labor farms paid $20,000 or more.

for 60 percent of FVH labor expenditures. The average FVH farm employer had labor expenditures of about $48,000 in 1982; about 9,100 or 12 percent of all FVH farm employers were deemed large, that is, they each had expenditures of $50,000 or more.

About one-third of all the persons hired in the 1982 COA were employed on FVH farms. Most workers employed on FVH farms were seasonal workers, that is, they were employed less than 150 days on the responding farm. The surprise in these top 10 states, ranked by the total labor expenditures of FVH farms, is the presence of states such as PA, NY, MI, and NJ. These northeastern states appear in the list because of their relatively large numbers of regular or more than 150 day workers who are paid relatively high wages; none except Michigan has a substantial number of seasonal workers. Note that, according to the 1982 COA, Pennsylvania FVH employers hired more seasonal workers directly than FVH employers in Texas, although Texas employers may have hired seasonal workers through labor contractors.

Fruit employment and expenditures are even more concentrated on a a handful of farms in a few states. Once again, California dominates, with over half of all fruit labor expenditures and employment. However, Oregon and Washington include one-quarter of all seasonal fruit workers while Hawaii includes 10 percent of all regular fruit workers, reflecting the relatively short harvests of the Pacific Northwest and the year-round nature of Hawaii agriculture.

The COA generates the most detailed farm labor data, and thus can be used to determine e.g. how much lettuce farms in a particular county spent for labor. Since the COA no longer reports the volume of hiring, its primary virtue will be using the labor expenditure data to indicate each state or county's relative volume of paid employment. Such labor expenditure data divided by an indicator of hourly earnings will be a useful indicator of each state or county's share of paid employment on farms.

Quarterly Agricultural Labor Survey (QALS)

The QALS data, published four times each year by USDA in Farm Labor, is one of the few farm labor data series used in regulatory proceedings. The QALS data on the hourly earnings of field and livestock workers are used by the U.S. Department of Labor to determine the Adverse Effect Wage Rate (AEWR); the

AEWR under the H-2A program is the wage rate that employers requesting temporary alien workers must offer to both U.S. and alien workers. QALS data are also used in many farm labor market studies to indicate e.g. how farmworker employment has changed over time and how farm wages change relative to nonfarm wages.

USDA has conducted a farm employment survey since 1910. Between 1910 and 1974, USDA relied on "volunteers" to report their farm employment and wages each month. In 1974, the QALS was converted to a probability survey and conducted quarterly, hence QALS. Budget cuts in the early 1980s led to several years of limited data collection, but since mid-1984 the QALS has been restored to a quarterly survey.

The National Agricultural Statistics Service (NASS) conducts the QALS survey of farm employers in January, April, July, and October to obtain information on farm employment, hours worked, and wages paid. The survey was conducted four times in five regions (California, Florida, Hawaii, Texas-Oklahoma, and Arizona-New Mexico) and three times (April, July, and October) in the 14 remaining regions, for a total of 18 regions from 1984 through 1988; however, because of the data requirements of immigration reform, QALS will be conducted 4 times annually in all regions beginning in January 1989. Farm employment, hours, and wages data are published for these 18 regions four times each year about one month after the survey is conducted in the USDA's Farm Labor publication.

The QALS sample is drawn from the list of farm employers maintained by USDA to conduct most of its periodic surveys, such as the number of livestock on farms and the acreages of various crops. The USDA list of sample farm employers is drawn from two sources or frames: a list frame or a list of farms known or likely to employ farm labor because of their size or major commodity and an area frame of about 16,000 land units in the United States. The area frame is used to supplement or correct for omissions from the list frame, and the master sample list is then corrected for duplication. For example, about 1,200 farms are interviewed in California; 1,000 are drawn from the list frame, and 200 from the area frame.

NASS is very reluctant to divulge specific information on its sampling procedures, but the master list contacted to conduct the survey each quarter includes about 13,500 names. This is a stratified sample, stratified (presumably in each state) by the number of workers employed or wages paid. About 50 percent of the sample is rotated each quarter, so that, over four quarterly surveys, about 27,000 farms might be contacted. Of the sample farms contacted, about 8,000 to 8,500 typically report that they hired labor during the survey week, and 4,000 to 4,500 report that they hired field and livestock workers during the survey week. Sampling details like these raise questions about this most widely used farm labor data; why do only one-third of the farms on a list of farms "likely to employ labor" actually report hiring field and livestock workers, and is a sample of less than 5,000 farm employers who actually hire field and livestock workers adequate to get reliable regional wage data?

QALS defines a farm as a place which sells farm products worth $1,000 or more annually, as in the COA. Farmworkers include all of the persons employed on crop (01) and livestock (02) farms, except those who regularly do nonfarm work such as carpentry, domestic household work, or do work which materially changes the form of the product, so that packing peaches into boxes on the farm is farmwork but canning them is not.

The QALS survey obtains employment data on farm operators, unpaid workers, and hired workers. The operator is asked how many hours he and each of his partners worked during the survey week (the Sunday through Saturday which includes the 12th day of the month). Operators are also asked about family or neighbor workers who did at least five hours of work during the survey week without pay. This unpaid group is heterogeneous: it includes children who help with farm chores if they receive an allowance but not a wage based on time worked, and prisoners, and workers on Indian and religious farms. These questions generate employment data, such as during the week of October 7-13, 1984, an estimated 1.5 million farm operators (including partners) averaged 43 hours of work on their farms and almost 600,000 unpaid workers averaged 36 hours of work.

Hired workers include all persons on the farm's payroll during the survey week, including paid family members, part-time workers, and hired managers. This means that at least some of the hired workers reported in Farm Labor are not stereotypical strangers-in-the-fields; they include relatives, bookkeepers and managers as well as field and livestock workers.

Hired workers are grouped by the type of work they do (field, livestock, supervisor, and other such as bookkeeper) and the method by which they are paid (hourly, piecerate, or other such as salary). The QALS groups workers by what they were hired to do, not what they actually did, so a supervisor who actually thinned lettuce during the survey week is classified as a supervisor. Workers doing a variety of tasks on the farm are classified by the type of farm that employs them, so that a general laborer on a crop farm is a fieldworker and a general laborer on a livestock farm is a livestock worker. Even though data is collected by the type of work done and the method of pay, the number of fieldworkers or the number of piecerate workers is not reported: only the total number of hired workers expected to be employed on the sample farm for more and less than 150 days during the year. In July 1987, USDA estimated that 706,000 hired workers would be employed more than 150 days on the responding farm and 564,000 would be employed on responding farms less than 150 days. Unlike the COA, where seasonal workers outnumber regular workers 4 to 1 over the year, in the QALS snapshot survey, regular workers outnumber seasonal workers about 2 to 1.

The most visible item from the Farm Labor survey is the wage rate paid to hired workers. Farm Labor reports eight hourly wages for hired workers, but the survey never asks farmers directly to report the average hourly or piecerate wage paid. Instead, farm employers are asked to report the number of workers who are paid in a certain manner, say 10 fieldworkers paid hourly wages, and then their total hours worked during the survey week (say 300) and their total gross wages (say $1,500). Farm Labor then divides gross wages by total hours to report that fieldworkers who were paid hourly wages average $5.

This indirect method of computing hourly wages has strengths and weaknesses. Reporting gross wages and hours permits the calculation of a single hourly wage despite the diversity of agricultural wage systems. However, this strength is also a weakness, especially if the hourly wage becomes a minimum or base wage in the H-2A regulatory process, because dividing gross wages by total hours "weights" the actual wages paid by hours worked. For example, if a farm has 10 irrigators who each work 60 hours during the survey period for $4 hourly, and 30 harvesters who each work 25 hours for $6 hourly, the average fieldworker wage is gross earnings ($2,400 + 4,500) divided by gross hours (600 + 750), or $6,900 ÷ 1350 hours = $5.11. Even though there are three times more harvesters than irrigators, the average wage of the sample farm is near the midpoint because the irrigators work more hours. The $5.11 wage is not the wage that either irrigators or harvest workers would obtain, even though employers who apply for H-2A workers are required to develop detailed job descriptions and offer a wage for such a detailed job.

Gross wages are divided by worker hours to generate average hourly wages for five types of workers (field, livestock, field and livestock combined, supervisory, and other) and three methods of pay (hourly, piecerate, and other). Gross wages are the gross earnings which should appear on an employee's pay stub; that is, gross wages exclude employer-paid payroll taxes and the costs of any fringe benefits provided to workers. Although NASS does not report any problems with this imputed wage survey procedure, its interviewer manual does note that it may be difficult for employers to easily report gross wages and hours, especially if their payroll period does not coincide with the survey week (e.g. it is biweekly or Wednesday to Wednesday). A second problem is the accuracy of total hours worked; Fair Labor Standards Act investigations show that hours worked, especially for piecerate workers, are often under reported in employer records.

The QALS also collects data on fringe benefits. All hired workers must be assigned to one of six benefit categories, with benefits ranked in a pre-assigned order. Employers are asked "how many workers receive or will receive"--although not necessarily during the survey week--(1) housing and meals; (2) housing only; (3) meals only; (4) cash bonuses; (5) other benefits such as health insurance; and (6) cash wages only. Workers can be assigned to only one of these six benefit categories, so e.g., many unionized California farmworkers will appear in the fifth category. This benefit question gives no guidance on the proportion of hired workers receiving any specific benefit except the first.

The QALS also asks operators about the value of their farm sales during the previous year, and whether field crops, FVH, or livestock and poultry contributed most of the farm's sales. Hired worker data by sales class began to be published in 1988.

QALS also collects information on agricultural service workers, who are workers employed by firms that provide soil preparation services (071), crop services such as combining grains or ginning cotton on the farm (072), veterinary (0741) and other livestock (0751) services, and farm labor and management services (076). QALS asks farm employers in all states except California and Florida to report the type of agricultural service and the number of service workers employed on their farm during the survey week. In California and Florida, QALS conducts a survey of agricultural service firms which generates service worker data comparable to that obtained from farm employers. Florida and California state offices maintain a list of agricultural service firms, and interviewers try to obtain the name of each firm used on a sample farm to determine if it is on the statewide list. A sample of agricultural service firms is contacted during each quarterly survey. These service firms are asked the same employment and wages questions as farm operators, e.g. they report workers, gross wages, and total hours by type of worker and method of pay.

In 1988, the QALS survey form was revised to help USDA to calculate employment in Seasonal Agricultural Services (SAS); SAS are the commodities (perishable crops) and activities (fieldwork) which received special treatment under the Immigration Reform and Control Act (IRCA) of 1986. The revised QALS form asks farmers in the sample to report, for workers hired directly and workers hired through agricultural service firms, the number of workers doing fieldwork (irrigating, pruning, harvesting) and the total man-days (four hours or more) worked in a list of commodities such as corn, cotton, fruits and nuts, nursery crops, and vegetables and potatoes. USDA has the responsibility for determining the number of man-days worked in SAS in FY 1989, but since it requests man-day data for only four weeks annually, USDA will have to assert that the four survey weeks are representative or estimate the man-days worked in crops such as raisin grapes which have their peak labor needs between survey periods (raisin grapes are harvested in September, between the mid-July and mid-October surveys).

IRCA also requires estimates of how many more workers would have been needed to prevent crop losses that were caused by labor shortages, and USDA added questions to the QALS which ask sample farmers if they had crop losses because of labor shortages and how many (additional) workers and man-days would have been needed to prevent these losses. Unlike the man-days worked questions, which refer only to the survey week, the crop loss questions refer to the past 12 months. USDA also added questions to determine if the "need" for SAS fieldworkers will change over the next 12 months, because of mechanization or different hiring practices. USDA will use this data to determine how many man-days were worked in SAS last year, how many were needed to prevent crop losses, and how many man-days are likely to be needed next year.

IRCA may revive interest in Farm Labor's employment estimates, which have traditionally attracted less interest than the wage estimates. Annual average wages in 1987 averaged $4.87

hourly for all hired workers, $4.69 for fieldworkers, $4.57 for field and livestock workers combined and $5.81 for piecerate workers. Piecerate hourly wages are usually 20 to 30 percent higher than the wages of workers paid on an hourly basis, and hourly wages are higher in California and the Pacific states than in the southeastern states (except Florida).

The April 1988 report includes some new cross-tabulations of data; Farm Labor reported that almost 1 million hired workers were employed, and 70 percent were regular or more than 150-day workers on responding farms. The average hourly wage of all hired workers was $5, and they averaged 40 hours of work during the survey week. The new data published in Farm Labor report that wages are highest in other (fruit and vegetable) crops and lowest in field crops, and that farms with sales of $250,000 or more pay the highest wages. In April 1988, these large farms accounted for 53 percent of all hired workers; in another cross-tabulation, Farm Labor reported that 41 percent of all hired workers were on farms with 11 or more hired workers.

Migrants are most likely to be fieldworkers, and the hourly earnings of field workers who were paid on an hourly basis are presented in Table 2.4 for the July survey weeks between 1980 and 1987. Hourly earnings rose 36 percent over this period across the U.S., from $3.38 to $4.60, but the increase was very uneven across regions. For example, between 1980 and 1987, hourly earnings increased only 18 percent in Iowa and Missouri, but over 60 percent in the northeastern states and Hawaii.

Year-to-year changes in hourly earnings are erratic, indicating a high variance in the averages estimated. High variances are indicated for the Mountain states[1]; in Idaho, Montana, and Wyoming, hourly earnings rose 29 percent between 1980 and 1987, but year-to-year changes were a 12 percent drop then a 34 percent increase, followed by a 10 percent drop, an 11 percent increase, a 23 percent increase, and in 1986-87 an 11 percent drop. Annual averages even out some of these fluctuations, but regulatory proposals to make the hourly earnings of reported in Farm Labor the minimum wage that farmers must offer when they request H-2A alien workers might be expected to exhibit some of these year-to-year fluctuations.

Farm Labor's employment estimates exhibit similar fluctuations. Farm Labor reported that the annual average hired employment was about 1 million between 1985 and 1987, and that about two-thirds

Table 2.4

Fieldworker Wages reported by Farm Labor for July: 1980-1987

Region	Average hourly wages of workers paid hourly wages							Percentage change in hourly wages						
	1980	1982	1983	1984	1985	1986	1987	1980-87	1980-82	1982-83	1983-84	1984-85	1985-86	1986-87
Northeast I	2.80	3.59	3.15	3.36	3.79	4.31	4.59	64%	28%	-12%	7%	13%	14%	6%
Northeast 2	3.14	3.62	3.83	3.84	3.69	4.20	5.13	63%	15%	6%	0%	-4%	14%	22%
Appalachia 1	2.88	3.47	3.37	3.46	3.50	3.91	3.84	33%	20%	-3%	3%	1%	12%	-2%
Appalachia 2	2.97	3.33	3.20	3.48	3.58	3.28	3.69	24%	12%	-4%	9%	3%	-8%	13%
Southeast	2.80	3.03	3.24	3.04	3.29	3.62	3.63	30%	8%	7%	-6%	8%	10%	0%
Florida	3.66	3.90	4.10	4.09	4.19	4.48	4.51	23%	7%	5%	0%	2%	7%	1%
Lake	3.23	3.47	3.21	3.72	3.67	4.38	4.51	40%	7%	-7%	16%	-1%	19%	3%
Cornbelt 1	3.51	3.80	3.80	3.92	4.06	4.22	4.77	36%	8%	0%	3%	4%	4%	13%
Cornbelt 2	3.38	3.69	3.92	3.69	4.20	4.34	4.00	18%	9%	6%	-6%	14%	3%	-8%
Delta	3.10	3.45	3.62	3.63	3.74	3.75	3.83	24%	11%	5%	0%	3%	0%	2%
North Plains	3.49	4.04	4.14	4.03	4.47	4.64	4.35	25%	16%	2%	-3%	11%	4%	-6%
South Plains	3.08	3.89	3.86	3.88	3.68	4.27	3.90	27%	26%	-1%	1%	-5%	16%	-9%
Mountain 1	3.13	2.76	3.70	3.33	3.68	4.51	4.03	29%	-12%	34%	-10%	11%	23%	-11%
Mountain 2	3.21	3.22	3.48	3.75	3.66	4.47	4.31	34%	0%	8%	8%	-2%	22%	-4%
Mountain 3	3.20	3.69	4.07	3.85	3.76	3.99	4.07	27%	15%	10%	-5%	-2%	6%	2%
Pacific	3.58	4.21	4.48	4.31	4.43	4.15	4.90	37%	18%	6%	-4%	3%	-6%	18%
California	4.12	4.69	4.58	4.88	4.83	5.22	5.54	34%	14%	-2%	7%	-1%	8%	6%
Hawaii	4.78	5.88	5.92	6.13	6.31	6.40	7.93	66%	23%	1%	4%	3%	1%	24%
United States	3.38	3.83	3.91	3.93	4.07	4.39	4.60	36%	13%	2%	1%	4%	8%	5%

Source: USDA, Farm Labor

Fieldworkers are nonsupervisory workers engaged in the the preparation, planting, caring, and harvesting of crops and harvesting of crops. Hourly wages are calculated by grouping all the fieldworkers who are paid hourly wages (say 100), reporting the total hours worked during the survey week by all of these workers (6 per day for 5 days is 3000), and listing the gross wages paid to these workers ($12,000). This calculation procedure yields an average fieldworker wage for each reporting farm; note that the $4 average hourly wage on one farm in this example may reflect the presence of equal groups of $3 and $5 per hour workers. This calculated wage procedure weights most the wages paid to workers employed the most hours during the survey week.

Northeast I: CT,ME, MA, NH, NY, RL, VT; Northeast 2: DE, MD, NJ,PA; Appalachian 1: NC, VA; Appalachian 2: KY, TN, WV; Southeast: AL, GA, SC; Lake: MI, MN, WI; Cornbelt: IL, IN, OH; Cornbelt 2: IA, MO; Delta: AR, LA, MS; Northern Plains: KS, NE, ND, SD; Southern Plains: OK, TX; Mountain 1: ID, MT, WY; Mountain 2: CO, NV, UT; Mountain 3: AZ, NM; Pacific: OR, v of the hired workers were regular or employed on the responding farm 150 days or more (Table 2.5). Hired worker employment is concentrated in a few areas: California had 16 percent of the average number of hired workers in 1987, followed by the Lake States of Wisconsin, Minnesota, and Michigan and then the Cornbelt States of Illinois, Indiana, and Ohio. Over the period 1985-87, hired worker employment decreased slightly, but it decreased 29 percent in the Appalachian states of Kentucky, Tennessee, and West Virginia, 22 percent in the Southeast, and over 10 percent in a broad area of the Midwest. As with hourly earnings, some of these year-to-year changes in employment probably reflect the variance in the estimate, not a change in actual farmworker employment.

The QALS is a source of timely farm wage and employment data with a long history and established usage in regulatory affairs. The data reported in Farm Labor are those normally used when comparing trends in farm and nonfarm wages or employment. The difficulties with Farm Labor data are twofold: first, too little information is released on the reliability of the sample data, but the erratic swings in wages from quarter-to-quarter suggest that the wage and employment data on hired workers is not reliable, especially for some regions. Second, as a regulatory instrument, the gross wages divided by total hours method of calculating hourly wages may not be generating the hourly wage figure that a typical farmer requesting H-2A alien workers is likely to offer to domestic and alien workers, especially if the workers requested are "supplemental" to regular or year-round workers who might be expected to work long hours. If farmers normally have to pay a premium wage in order to attract supplemental workers, then using a Farm Labor wage which is influenced most by the (regular) workers employed the most hours may result in too low a wage for supplemental seasonal workers, making it difficult to recruit American workers.

The Farm Costs and Returns Survey

The USDA's Farm Costs and Returns Survey obtains data on farmers' labor expenditures from a sample of about 24,000 crop and livestock farms. Labor expenditure data is reported for each state and for ten multi-state farm production regions by the USDA's Economic Research Service (ERS) and USDA's National

Table 2.5

Hired Workers on Farms as Reported in Farm Labor: 1985-1987

Region	Annual Average Employment: 1985						1986			Annual Average Employment: 1987						All Hired	
	All Hired (000)	Per Dist	Regular (000)	Per Dist	Seasonal (000)	Per Dist	All Hired (000)	Regular (000)	Seasonal (000)	All Hired (000)	Per Dist	Regular (000)	Per Dist	Seasonal (000)	Per Dist	Per Change 85-86	Per Change 85-87
Northeast 1	60	5%	40	6%	20	5%	62	41	20	61	6%	42	6%	19	5%	3%	2%
Northeast 2	52	5%	36	5%	16	4%	54	39	15	56	5%	41	6%	15	4%	3%	7%
Applachia 2	72	7%	35	5%	37	9%	57	29	28	51	5%	26	4%	25	7%	(21%)	(29%)
Applachia 1	35	3%	16	2%	19	4%	45	22	23	36	3%	16	2%	19	5%	26%	1%
Southeast	67	6%	35	5%	32	7%	46	31	16	52	5%	28	4%	24	6%	(31%)	(22%)
Florida	50	5%	41	6%	9	2%	56	43	11	54	5%	43	6%	11	3%	13%	8%
Lake	108	10%	59	9%	48	11%	97	55	42	99	9%	56	8%	43	11%	(10%)	(8%)
Cornbelt 1	87	8%	50	8%	37	9%	88	57	31	83	8%	51	8%	32	8%	2%	(5%)
Cornbelt 2	49	4%	24	4%	25	6%	50	26	24	44	4%	25	4%	19	5%	1%	(11%)
Delta	57	5%	38	6%	19	4%	51	37	13	49	5%	33	5%	16	4%	(11%)	(14%)
North Plain	59	5%	36	5%	24	5%	43	26	17	50	5%	33	5%	17	5%	(28%)	(16%)
South Plain	81	7%	57	9%	23	5%	72	46	25	72	7%	70	10%	25	7%	(11%)	(11%)
Mountain 1	30	3%	18	3%	12	3%	29	17	11	31	3%	18	3%	13	3%	(4%)	2%
Mountain 2	27	2%	19	3%	8	2%	26	17	8	25	2%	15	2%	10	3%	(5%)	(7%)
Mountain 3	18	2%	14	2%	4	1%	17	13	2	21	2%	17	2%	7	2%	(8%)	12%
Pacific	60	5%	23	3%	37	9%	62	31	30	63	6%	30	4%	33	9%	3%	6%
California	175	16%	117	18%	58	13%	167	126	42	172	16%	122	18%	50	13%	(5%)	(2%)
Hawaii	10	1%	9	1%	1	0%	10	9	1	11	1%	10	1%	1	0%	(5%)	5%
United State	1,099	100%	665	100%	434	100%	1,039	673	367	1,051	100%	675	100%	376	100%	(5%)	(4%)

Source: USDA, Farm Labor, Quaterly: Data based on about 8500 responses in the 14,000 farm sample.

Hired workers are persons who were paid cash wages to work on the sample farm during the week which includes the 12th of the month;

Regular workers are expected to be employed 150 or more days on

the responding farm; seasonal workers 149 days or less on the sample farm.

Northeast I: CT, ME, NH, RI, VT; Northeast 2: DE, MD, NJ, PA; Appalachian I: NC, VA; Appalachian 2: KY, TN, WV; Southeast: AL, GA, SC; Lake: MI, MN, WI; Cornbelt I: IL, IN, OH; Cornbelt 2: IA, MO; Delta: AR, LA, MS; Northern Plains: KS, NE, ND, SD; Southern Plains : OK, TX; Mountain I: ID, MT, WY; Mountain 2: CO, NV, UT; Mountain 3: AZ, NM; Pacific: OR, WA.

43

Agricultural Statistics Service (NASS). The ERS adjusts the labor expenditure data using Census of Agriculture and Social Security Administration data before publishing it, while NASS makes no adjustments to the sample data before publishing it.

The labor expenditure data collected includes more items than are reported in the COA or UI data. Labor expenditures are the cash wages, bonuses, in-kind and fringe benefits, and the employer share of Social Security taxes. Contract labor expenses, which are reported separately, include expenditures for combining, plowing, harvesting, and pruning done on a contract basis by a labor contractor, crew leader, or a cooperative. NASS reports separately expenditures for voluntary health and pension plans, Social Security and other payroll taxes, and the cash value of other voluntary benefits such as worker housing and meals; ERS reports different perquisites expenditure items. The FCRS apparently obtains the peak number of employees during the past year, but this employment data is not reported.

Labor expenditures are separated into "cash labor expenses" and "perquisites" in the ERS Economic Indicators of the Farm Sector: State Financial Summary. Cash labor expenses include contract labor expenses, hired labor wages, and employer Social Security payments. ERS reported that in 1986 U.S. cash labor expenses were $9.2 billion and that perquisites were worth $0.6 billion; for California, cash labor expenses were $2.2 billion and perquisites were $0.1 billion. ERS estimates of farmers' labor expenditures were revised upward in 1986 when the estimation procedure was modified.

NASS reports the FCRS labor expenditure data for ten multi-state regions in its Farm Production Expenditures report. For each region, NASS reports the percent of sample farms that hired labor, their total expenditures for labor, total expenditures for labor, information on the distribution of labor expenditures by farm sales class e.g. in the Pacific states of California, Oregon and Washington, about 72 percent of all cash wages were paid by farms with sales of $100,000 or more in 1986. In the mid-1980s, about half of the FCRS sample farms reported that they hired labor, and between 5 and 15 percent of the labor expenditures reported by sample farms were paid to relatives of the operator.

The NASS Farm Production Expenditures report for 1986 estimates that labor expenditures were $10 billion, or about 9 percent of farmers' production costs. About 80 percent of these expenditures were cash wages, and almost 10 percent were expenditures for contract labor. NASS reports that farmers' payroll taxes were about 7 percent of labor expenditures, and the cash value of voluntary fringe benefits was about 6 percent; in the nonfarm sector, by contrast, payroll taxes are about 8 percent of total compensation and voluntary fringe benefits are 18 percent.

The FCRS is primarily a labor expenditure survey which obtains data to estimate what farmers spent on hired labor. Despite the inclusive nature of the labor expenditures reported, they can be used to distribute the volume of farmworker activity across states. Payroll taxes are relatively uniform, especially within states, so labor expenditure data can be divided by an hourly wage to obtain an hours worked indicator by state and region, similar to the method suggested to adjust COA data. The FCRS is an annual survey, and is thus capable of generating timely state shares of the volume farmworker employment activity.

BEA Farm Labor Estimates

The Bureau of Economic Analysis (BEA) division of the Department of Commerce makes annual estimates of farm employment and farm labor expenditures for states and counties. BEA does not collect any primary data; instead, BEA relies primarily on data from the annual FCRS. However, BEA adjusts this data differently than does ERS, and BEA makes estimates for both states and counties, while ERS makes estimates only for states.

BEA makes estimates for labor expenditures and employment. Labor expenditures include hired farm labor expenses, contract labor expenses, machine hire and custom work expenses, total farm wages and salaries, and other farm labor income. Hired labor expenses are similar to the COA labor expenditures concept; they include wages and salaries, payroll taxes, and fringe benefits. Total farm wages and salaries include cash wages and the value of in-kind fringe benefits (apparently excluding payroll taxes and employer contributions for insurance and other fringes). Other farm labor income consists of employer contributions for pension and insurance funds and for private workers compensation plans; however, the officers of corporate farms are considered to be hired workers by BEA, and other labor income includes directors fees for such farms.

BEA also estimates farm wage and salary employment and self-employment on farms. These employment estimates are based on the QALS, the COA, and the ES-202 UI data; they do not separate full and part-time workers. The employment estimates appear to be (low) average annual estimates for crop and livestock farmers e.g. for 1986, BEA reported 1 million wage and salary workers in U.S. agriculture and 160,000 in California, while California UI data indicate that average annual employment on crop and livestock farms was 228,000.

The BEA ranking of labor expenses has been very stable since 1979 and it is very similar to the 1982 COA (Table 2.6). The same states are included in the top ten states in both COA and BEA, but BEA hired labor expenses are higher than COA labor expenditures for 1982: BEA reports $10.4 billion, and COA reports $8.4 billion. California accounts for 20 percent of BEA's farm labor expenses, and California, Texas, and Florida account for one-third of U.S. farm labor expenditures.

The BEA estimates are the only ongoing adjustments of farm labor data which produce annual wage and employment indicators for states and counties. The adjustment procedures are not spelled out, but they may be the best adjustments possible to obtain timely state and county farm labor indicators.

HOUSEHOLD DATA

Household data are obtained from individuals, and households or individuals are the major source of data on the demographic characteristics of the population and workforce, employment and earnings, and migration across occupations, industries, or areas. Some household information does not change over time, such as sex and race, some changes in a predictable fashion (age), and other household characteristics such as employment status and earnings can be quite variable from month-to-month, especially for seasonal farmworkers. Thus, household questionnaires obtain employment and earnings information by asking about a particular reference week, month, or year.

Census of Population

The Census of Population (COP) has been conducted every 10 years since 1790. The COP collects a variety of demographic data on the age, sex, and race of individuals as well as economic data on

Table 2.6

BEA Estimates of Farm Labor Expenses: 1979-1986

State	Hired Farm Labor Expense 1979-$000-	Per Dist	Hired Farm Labor Expense 1982-$000-	Per Dist	Hired Farm Labor Expense 1984-$000-	Per Dist	Farm Wages & Salaries 1986-$000-	Per Dist
California	1,721,988	20%	2,121,933	20%	2,098,687	20%	1,630,078	20%
Texas	547,277	6%	640,915	6%	633,667	6%	466,515	6%
Florida	483,474	6%	581,252	6%	574,994	6%	466,463	6%
Washington	312,552	4%	383,563	4%	379,304	4%	302,563	4%
North Carolina	307,946	4%	330,040	3%	326,292	3%	230,251	3%
Wisconsin	264,808	3%	316,447	3%	312,898	3%	276,531	3%
Illinois	282,683	3%	304,216	3%	300,710	3%	225,375	3%
Iowa	284,948	3%	300,653	3%	297,163	3%	263,171	3%
New York	232,081	3%	288,269	3%	285,011	3%	233,725	3%
Pennsylvania	190,527	2%	254,785	2%	251,917	2%	206,352	3%
Top Ten States	4,628,284	54%	5,522,073	53%	5,460,643	53%	4,301,024	52%
United States	8,591,069	100%	10,419,364	100%	10,302,276	100%	8,243,000	100%

Source: Bureau of Economic Analysis, U.S. Department of Commerce; Farm Income and Expenditures series

BEA state and county estimates are based on FCRS data from ERS; ERS lumps labor expenditures and contract labor together, while BEA separates them

Hired farm labor expense is similar to the COA labor expenditures concept; it include cash wages paid to all wage and salary workers, employer payroll taxes and the value of any fringe benefits offered to workers

Farm wages and salaries include cash wages paid to hired workers, in-kind benefits for hired workers and the salaries received by officers of corporate farms; it excludes contract labor expenses their incomes, occupations, and industries of employment. The COP reports these demographic and economic data by state, county, and city. The COP is mailed to all households in April of the Census year (April 1980), and respondents are asked to report employment data for the last week in March and income data for the previous year (1979).

COP employment data is collected from a 19 percent sample of households. All persons 15 and older report whether they were employed, unemployed but looking for work, or not in the labor force, but employment data is published only for persons 16 and older. Respondents are to report only their chief employment-related activity during the last week of March, so that respondents with two jobs during the reference week describe the one at which they worked the longest. Persons not in the labor force and unemployed persons report on their last job held.

The 1980 COP grouped jobs into 503 occupations. There are 13 agricultural occupations, and they are divided into farm operators and managers; other non-managerial farm occupations; and related agricultural occupations. The four non-managerial occupational titles are farmworkers (479) and farmworker supervisors (477), marine life cultivation workers (483), and nursery workers (484). In most farm labor statistics and in administrative UI data, farm managers (475) and managers of horticultural specialty farms (476) are also considered to be "hired farmworkers".

The 1980 COP also asked respondents about their industry of employment, and the Census reported industry of employment data for 231 industries. However, instead of reporting all 50 agricultural SIC codes, the Census reported only four: crops (01), livestock (02), farm agricultural services (07*), and horticultural services (078).

COP employment data divides respondents into several groups: self-employed, unpaid family workers who worked at least 15 hours in the reference week, and wage and salary workers. Wage and salary income during the previous year is all earnings, bonuses, and piece rate payments before deductions. However, workers employed in several occupations or industries report only their total earnings, so that it is not possible to link an individual's COP earnings with a particular industry or occupation. Since many farmworkers also have nonfarm jobs during the year, it is not possible to divide their earnings into farm and nonfarm sources in the COP.

The Census of Population also collects data on the number of weeks worked during the previous year, hours worked during the reference week, and usual hours worked per week during the past year. However, this weeks and hours data refer to all jobs, and thus it is not possible to isolate weeks and hours by occupation or industry.

The Census of Population is not often used to enumerate or distribute farmworkers across states for several reasons. Many ·farmworkers change occupational titles during the year so the earnings, weeks, and hours data attributed to farmworkers include an unknown amount of nonfarm employment and the nonfarm data include an unknown amount of farm work. A more serious problem with using COP data to analyze farmworkers is that a COP farmworker must have done farmwork during the March reference week, and March is a month of relatively little seasonal farmworker employment in most states. An analysis by Whitener of farmworkers employed in March shows that such workers are much more likely than farmworkers not employed in March to be year-round livestock workers who depend on farmwork for most of their earnings.

An additional problem with using Census of Population data is that migrant farmworkers are not identified. One could classify the "farmworkers" included in the Census by weeks worked and earnings, but both of these indicators refer to weeks and earnings from both farm and nonfarm jobs.

The 1980 COP reported 875,000 farmworkers and 1.1 million farmers (Table 2.7). According to the COP, both farmworkers and farmers were mostly White; 65 percent of the farmworkers and 97 percent of the farmers were White in the 1980 COP. The presence of year-round White farmworkers is evident by the inclusion of Wisconsin, Iowa, Minnesota and Illinois among the 10 states with the most farmworkers; these states did not include even 1 percent of all minority farmworkers. According to the COP, Hispanic farmworkers are heavily concentrated in California (58 percent of Hispanic farmworkers in the COP) and Texas (18 percent).

Current Population Survey

The Current Population Survey (CPS) is the source of monthly employment information, for example, the CPS is the source of the monthly unemployment rate. The CPS is based on a sample of

50

Table 2.7
Farmworkers and Farmers in the 1980 Census of Population

Farmworkers

State	All Farmworkers	Per Dist	White	Total Minority	Per Dist	Black	Hispanic	Per Dist	Female farmworkers	Per Dist
California	150,116	17%	34,493	115,623	38%	2,147	106,115	56%	33,407	18%
Texas	65,276	7%	26,778	38,498	13%	3,990	34,267	18%	10,357	5%
Florida	51,824	6%	19,142	36,682	12%	20,481	11,634	6%	14,481	8%
Wisconsin	34,688	4%	34,278	410		47	214		11,261	6%
Iowa	30,022	3%	29,831	191		22	98		8,354	4%
North Carolina	29,407	3%	14,708	14,699	5%	13,433	416		8,021	4%
New York	27,800	3%	26,046	1,754	1%	902	628		6,198	3%
Minnesota	26,111	3%	25,848	263		9	137		7,754	4%
Illinois	25,522	3%	24,439	1,083		283	686		5,156	3%
Washington	23,381	3%	14,967	8,414	3%	155	7,344	4%	5,466	3%
Top Ten States	464,147	53%	250,530	217,617	71%	41,469	161,539	85%	110,455	58%
United States	874,784	100%	568,453	306,838	100%	92,600	189,263	100%	189,659	100%

Farmers

State	Total Farmers	Per Dist	White	Total Minority	Per Dist	Black	Hispanic
Iowa	86,591	8%	86,433	158		10	109
Texas	70,094	6%	65,780	4,314	12%	844	3,263
Minnesota	69,692	6%	69,552	140		5	67
Wisconsin	67,478	6%	67,325	153		11	78
Illinois	61,664	6%	61,416	248	1%	69	117
Missouri	53,875	5%	53,580	295	1%	71	150
Nebraska	47,951	4%	47,788	163		12	71
Ohio	39,955	4%	39,712	243	1%	75	135
California	37,509	3%	31,646	5,863	16%	188	2,938
Indiana	37,144	3%	36,999	145		24	83
Top Ten States	571,953	52%	560,231	11,722	33%	1,309	7,011
United States	1,101,060	100%	1,065,022	36,038	100%	15,814	11,520

Source: Census of Population, 1980; employment data is based on a 19 percent sample of households
Occupation is based on the responding individual's chief job activity during the last week in March, 1980
Farmers includes all farm operators and managers except hort specialty farmers

about 56,000 households throughout the U.S. The CPS is a household survey--if a family moves, the new family living in the sample household is interviewed. The CPS survey is constructed so that all U.S. housing units have an equal probability of being included.

Employment data are collected by the CPS each month and reported monthly by the Bureau of Labor Statistics in Employment and Earnings. This employment data is generated from questions that are very similar to those used in the Census of Population, for example, individuals are asked to report the job at which they worked the longest during the week containing the 12th of the month. Employment data is reported by agricultural industry (crops, livestock, and agricultural services) and for 14 farmworker occupations. However, as in the COP, employment, earnings, and hours refer to all jobs held during the survey week, but industry and occupation refer to the job at which the respondent worked the longest. Some of the monthly CPS data for farmworkers is grouped with farm operators and managers; e.g., the CPS reported that 3 million persons were employed in agriculture in June 1988, that 1.6 million were wage and salary workers, and that 10 percent of the farm wage and salary workers were unemployed.

The Hired Farm Working Force (HFWF) survey is a USDA-ERS designed supplement to the regular monthly CPS. The HFWF survey is conducted every other year in December, when migrant farmworkers have presumably returned to their homes, so that the HFWF enumerates migrants where they live, not where they work. The HFWF asks questions about the preceding calendar year; all of the 59,500 households in the CPS are asked the screening question (In 1985 did anyone in this household do any farmwork for cash wages or salary even for one day?), and then the 1,500 households which respond yes are asked a series of questions about their farm employment and earnings; these households included 1,880 farmworkers in 1983, 1,771 in 1985, and 1,647 in 1987.

The basic CPS questionnaire obtains demographic and survey week employment data, and the HFWF supplementary questions ask for farm days worked and farm earnings for each month of the past year. All farmworkers were asked if they left home and stayed away overnight to harvest crops or do other farmwork for cash wages. A "yes" response to this question is followed by a question to determine if the overnight stay was in a different county. Thus, about 1,500 of the 59,500 CPS households included at least one person who reported doing farmwork for wages sometime during the year (2.5 percent), and about 120 of these 1,500 farmworker households (8 percent) included a person who stayed away overnight in a different county to do farmwork in 1983.

The CPS-HFWF survey defines a migrant farmworker as a person who stays away from home overnight in a different county to do farmwork for wages. Once a person is defined as a migrant farmworker in the HFWF, all of that person's farmwork during the year is considered migratory farmwork, so that if the worker did one day of migratory farmwork and 100 days of local farmwork, the HFWF reports 101 days of migratory farmwork. Migrant farmworker households are found in about 20 states, and the HFWF expands the sample migrants in these states to generate an estimate of the number of migrant farmworkers in 10 multi-state regions.

The CPS-HFWF survey data has been reported in a variety of ways over the past decade, but the data for selected years between 1976 and 1983 in Table 2.8 exhibit a remarkable consistency; about 2.5 to 2.7 million persons did farmwork for wages sometime during the year, and 190,000 to 225,000 were migrants. However, the small migrant sample (about 120 migrant households throughout the United States) makes the regional distribution of migrants suspect; e.g. sharp 1976 to 1983 changes in the number of migrants, such as the jump from 1,000 in the Northeast to 7000 or from 9000 in the Pacific states to 55,000, probably reflect problems with the sample size and sampling procedures, not changes in the distribution of migrants.

Each HFWF survey obtains data on days worked and annual earnings from farm and nonfarm work. In the past, migrants have been asked how far they travelled to do farmwork, whether they were paid hourly or piecerate wages, and the primary commodity produced on the farm on which the migrant was employed longest.

The HFWF is the only national statistical survey that collects demographic and employment data on migrant farmworkers. In 1985, the HFWF reported that 159,000 of 2.5 million farmworkers or 6 percent of all farmworkers were migrants. Migrants ranged from 2 percent of all farmworkers in the northeastern states in December to 15 percent in the southeastern states. Migrants were reported to be 72 percent White, 9 percent Black, and 19 percent Hispanic; 57 percent of all migrants did less than 75 days of

Table 2.8
Farmworkers and Migrants from the December CPS: 1976-1983

CPS-HFWF-1976

Region	All Hired Farmworkers	Migrant Farmworkers	Migrant Per of All
Northeast	77,000	1,000	1%
Lake States	94,000	3,000	3%
Corn Belt	110,000	12,000	11%
Northern Plains	670,000	25,000	4%
Appalachian	446,000	18,000	4%
Southeast	441,000	83,000	19%
Delta States	249,000	11,000	4%
Southern Plains	156,000	19,000	12%
Mountain	320,000	31,000	10%
Pacific	204,000	9,000	4%
United States	2,767,000	213,000	8%

CPS-HFWF-1983

Region	All Hired Farmworkers	Migrant Farmworkers	Migrant Per of All	Migrants who worked in 1983 Less than 75 days	75 to 149 days	150 or more days
Northeast	217,000	7,000	3%	7,000		1,000
Lake States	199,000	10,000	5%	4,000	3,000	2,000
Corn Belt	376,000	15,000	4%	11,000	1,000	3,000
Northern Plains	112,000	7,000	6%	4,000	1,000	2,000
Appalachian	381,000	13,000	3%	8,000	2,000	2,000
Southeast	296,000	77,000	26%	7,000	38,000	32,000
Delta States	96,000	6,000	6%	3,000	1,000	2,000
Southern Plains	194,000	19,000	10%	11,000	3,000	5,000
Mountain	189,000	18,000	10%	10,000	4,000	4,000
Pacific	534,000	55,000	10%	23,000	8,000	24,000
United States	2,595,000	226,000	9%	88,000	61,000	77,000

CPS-HFWF-1977

Federal Region	All Hired Farmworkers	Migrant Farmworkers	Migrant Per of All
1	75,000	1,000	1%
2	81,000		
3	153,000	4,000	3%
4	706,000	26,000	4%
5	400,000	18,000	5%
6	364,000	70,000	19%
7	277,000	14,000	5%
8	172,000	24,000	14%
9	301,000	27,000	9%
10	202,000	6,000	3%
U.S.	2,730,000	191,000	7%

CPS-HFWF-1979

Federal Region	All Hired Farmworkers	Migrant Farmworkers	Migrant Per of All
1	65,000	1,000	2%
2	85,000	8,000	9%
3	118,000	9,000	8%
4	608,000	28,000	5%
5	488,000	33,000	7%
6	403,000	53,000	13%
7	261,000	20,000	8%
8	107,000	15,000	14%
9	364,000	39,000	11%
10	152,000	11,000	7%
U.S.	2,652,000	217,000	8%

CPS-HFWF-1981

Federal Region	All Hired Farmworkers	Per Dist
1	60,000	2%
2	101,000	4%
3	119,000	5%
4	567,000	23%
5	380,000	15%
6	384,000	15%
7	269,000	11%
8	107,000	4%
9	334,000	13%
10	172,000	7%
U.S.	2,492,000	100%

Source: Current Population Survey, Supplemental questions attached to the December survey

In recent years, about 1500 households included 1900 persons who did farmwork for wages sometime during the year

Migrants crossed county lines and stayed away from home overnight to do farmwork for wages; there were130 sample migrants farmwork in 1985 and only 20 percent primarily did farmwork for wages in 1985. Migrants had average total earnings of $5,700, versus $5,800 for non-migrants.

The number and characteristics of migrant farmworkers located by the CPS has fluctuated since the mid-1970s. For example, the number of migrants was 115,000 in 1981 and 226,000 in 1983 and the percentage of White migrants rose from 45 percent in 1983 to 72 percent in 1985. The USDA's HFWF report notes that migrant worker estimates must be "interpreted cautiously because they are based on a small number of (sample) cases."

The HFWF survey was originally scheduled for December so that migrants could be interviewed at home. Changing migration patterns and the rise of illegal alien workers may be partially responsible for what are perceived to be "too few Hispanic" migrants estimated in the HFWF, although there is no alternative national statistical survey which produces a larger number.

The original interview-the-migrants-at-home-in-December strategy was based on the assumption that most migrants did farmwork out-of-state. However, the HFWF reports that most migrants did farmwork in the state where they were interviewed, suggesting that the distribution of the HFWF sample is a guide to both the residences and the workplaces of HFWF migrants. Unpublished data indicate that 49 percent of the sample migrants did farmwork in the state where they were enumerated in 1981, 78 percent in 1983, and 56 percent in 1985 (Table 2.9).

In 1981 and 1983, migrants were asked the longest distance they travelled to do farmwork for wages. Distance-travelled data also indicate that most migrants do not travel great distances: 43 percent travelled less than 400 miles as a migrant in 1981, and 29 percent travelled less than 400 miles in 1983 (Table 2.10). A 400 mile trip would keep most of the migrants based in southern Florida, south Texas, and southern California within their home states.

ADMINISTRATIVE DATA

Administrative data are generated when farmers pay unemployment insurance (UI) and social security taxes for their employees, when farmworkers report their occupation on Internal Revenue Service tax forms, and when farmworker assistance programs obtain data from the clients they serve. Administrative

Table 2.9

Where CPS-HFWF Migrants Worked: 1981, 1983, 1985

Migrants who worked*:	1981	Percent	1983	Percent	1985	Percent
Within state	44,000	38	162,000	72	70,000	45
In another state	58,000	50	58,000	26	88,000	57
In and out-of-state	13,000	11	13,000	6	17,000	11
All Migrants	115,000	100	226,000	100	155,000	100

*Migrants stayed away from home overnight in a different county to do farmwork for wages.

Source: Unpublished data from the CPS, December 1981, 1983, and 1985.

Table 2.10

Longest Distance Travelled as a Migrant Farmworker:
1981 and 1983

Migrants who travelled:	1981	Percent	1983	Percent
Less than 75 miles	33,000	29	29,000	13
75-199	16,000	14	36,000	16
200-399	12,000	10	35,000	15
400-499	3,000	3	5,000	2
500-999	25,000	22	52,000	23
1,000-1,499	16,000	14	18,000	8
1,500-1999	3,000	3	49,000	21
2,000 miles or more	7,000	6	3,000	1
All Migrants	115,000	100	226,000	100

Source: CPS-HFWF, December 1981 and 1983.

data are used widely to chart trends in nonfarm labor markets, for example, administrative data from nonfarm employers are used to chart trends in factory employment and hourly earnings.

Administrative data have not been used widely to chart trends in the farm labor market for several reasons. First, the coverage of farmworkers under programs such as UI is incomplete, so that administrative UI data paint only a partial picture of the farm labor market. Second, some of the administrative data collected on farmworkers is of very inconsistent quality, such as the Department of Labor's ES-223 In-Season Farm Labor Reports. Finally, client data may be seen as self-serving by assistance organizations requesting funds and, since every farmworker assistance program asserts that it serves only a fraction of its target population, it is hard to adjust client data to represent the true population. In addition, the distribution of clients across states probably reflects as much the distribution of assistance funds as it does the distribution of farmworkers, especially if each dollar spent generates about the same number of clients in each state.

Administrative data have problems, but they also have advantages. First, since administrative data are already collected, they are relatively cheap to obtain and manipulate. Second, administrative data may provide more complete coverage than statistical sources; the UI data, for example, are a "census" of all covered employers and their employees. Third, administrative data forms can be designed to collect precisely the data wanted e.g. migrant clinic intake forms can be designed to collect the precise demographic and employment data desired on clients.

This section reviews three administrative data sources. The first is the ES-202 or UI data which are compiled from the farm employers who are required to provide UI coverage to their employees. The second administrative data source is the ES-223 In-Seasonal Farm Labor Reporting System; these are reports submitted by Employment Service staff in various states and published by the U.S. Department of Labor. Finally, the Migrant Student Record Transfer System (MSRTS), which reports data on the enrollment of the children of migrant farmworkers, is described. This section concludes that the UI data are the most reliable administrative data because farm employment is being concentrated on fewer and larger farms, so that more and more farmworkers are being included in this basic administrative data source which is the source of most nonfarm employment and wage data.

ES-202 Employment and Wages Program or UI Data

The Unemployment Insurance (UI) program is a federal-state system enacted in 1938 to partially replace earnings lost due to unemployment. The federal government establishes minimum coverage levels (to determine which employers must cover their employees) and levies a tax on the first $7,000 earned by each covered worker to pay for the administration of the system. However, each state determines exactly what type of job loss constitutes unemployment and the level of benefits paid to unemployed workers.

The federal government mandated UI coverage of farmworkers employed by large farm employers in 1978. Since then, farms that paid cash wages of $20,000 or more for farm labor in any calendar quarter in the current or preceding year or which employed 10 or more workers on at least one day in each of 20 different weeks must provide UI coverage for their employees. This "20-10" rule results in coverage of only larger farm employers; coverage was estimated to be 40 percent of all farmworkers in the mid-1970s, but the continuing concentration of farm production on fewer and larger means that in the 1980s about 70 to 80 percent of all farmworkers are probably employed on farms which must provide UI coverage to their workers. Federal UI law defines a farm employer as an entity that hires workers to raise or harvest agricultural or horticultural products on a farm; repair and maintain equipment on a farm; handle, process, or package commodities if over half of the raw commodity was produced by the employer; gin cotton; or do housework on a farm operated for profit. Seven states--California, Maine, Minnesota, Rhode Island, Virginia, Texas, and Washington--require additional agricultural employers to provide UI coverage for their employees and several other states are considering requiring smaller farm employers to provide UI coverage to their workers.

The Employment and Wages Program generates data from the quarterly reports which employers are required to file with state UI programs when they pay their UI taxes. Covered employers are required to report all persons on the payroll for the payroll period which includes the 12th day of the month and total wages paid for the quarter. When obtaining their UI account numbers, employers provide information on their headquarters county of operation and their primary commodity or activity.

Selected employment and wage data are forwarded by the states to the Employment and Training Administration in Washington, which publishes this data annually in Employment and Wages. Data are published for each state at the two-digit SIC level, that is, for crop, livestock, and agricultural service employers, and are available for more detailed commodity groupings, so that the number of employers (reporting units), average monthly employment, and total farm wages can be determined by commodity for each state. The UI system tabulates reporting units, not employers, so that a single employer with separate crop and livestock operations which each employ 50 or more workers would be two reporting units in UI data. Similarly, separate operations in two counties should generate a reporting unit in each county.

The UI system defines employees as all persons on the payroll, so that the paid managers of an agricultural corporation, supervisors, office personnel, professional staff, and fieldworkers are all considered to be "farmworkers". Farm employers report only cash wages; not payroll taxes, bonuses or the cost of fringe benefits. Averages computed from UI data must be interpreted carefully: "employers" may represent several sub-units of one large farm; reported employment increases with the length of the payroll period because of turnover; and wages represent pay to both professionals and to fieldworkers.

The major advantage of UI data are that they represent a "census" of covered employers and workers. This means that as more farmworkers find employment on the larger farms which must provide UI coverage to their employees, the ES-202 data may become a standard source of farm labor data. The UI system obtains basic employment and earnings data by commodity and county; the only data element missing to generate hourly earnings is hours worked, so that adopting the nonfarm procedure of asking a sample of employers to report monthly the hours worked and gross earnings of groups of workers such as women and production workers would generate average weekly hours and hourly earnings by month for farmworkers in a fashion similar to that used in the QALS.

UI covered employment and wages are reported in Table 2.11 for employers producing fruits and nuts, vegetables, and horticultural specialties. The concentration of employment and wages in California is evident; what is surprising is the presence of states such as Pennsylvania, New York, Illinois, and New Jersey

Table 2.11
Farm Employment and Wages Covered by Unemployment Insurance in 1985

State	FVH Employers	Per Dist	Ave Annual Employment	Per Dist	Annual Wages($)	Per Dist	FVH as a Percent of all UI Crop Employment Employers	Employment	Wages
California	11,737	54%	150,769	42%	1,631,636,656	44%	70%	76%	74%
Florida	1,652	8%	50,664	14%	504,720,652	14%	94%	92%	84%
Washington	887	4%	20,949	6%	143,919,447	4%	70%	78%	71%
Texas	301	1%	12,249	3%	122,083,794	3%	41%	72%	68%
Pennsylvania	429	2%	10,656	3%	118,894,156	3%	91%	96%	97%
New York	712	3%	9,214	3%	102,899,655	3%	88%	93%	93%
Oregon	422	2%	10,133	3%	85,774,066	2%	63%	74%	71%
Illinois	268	1%	9,285	3%	84,666,295	2%	51%	77%	69%
New Jersey	622	3%	7,036	2%	78,981,995	2%	95%	98%	98%
Michigan	382	2%	7,348	2%	72,993,877	2%	61%	71%	71%
Top Ten States	17,411	80%	288,302	80%	2,946,570,593	80%	62%	72%	70%
United States	21,838	100%	361,901	100%	3,698,417,050	100%			

Source: ES-202 Employment and Wages Program; 40 to 60 percent of all fws are reported by employers subject to UI

Most states have Federal Unemployment Tax Act (FUTA) coverage of farm employers; farm employers who pay quarterly cash wages of $20,000 or more or employ 10 or more workers for at least one day in each of 20 different weeks must cover their farmworkers (20-10 rule). CA, ME, MN, RI, TX, and WA have more inclusive farmworker coverage.

Employers are farms which employ 50 or more workers; this column is the number of ers each quarter divided by four.

Ave annual employment is the employer-reported count of all workers on the payroll for the pay period which includes the 12th day of the month summed over 12 months and divided by 12; this column is the average of four quarterly reports.

not normally considered large employers of the workers associated with migrants and illegal immigration. The presence of such states in UI data may reflect the concentration of production on large farms which must provide UI coverage to their workers. It is interesting to note that, in these states, FVH employers represent 70 to 90 percent of all UI-covered crop employment e.g. in California, grain, field crop, and general crop farms pay just 26 percent of UI-covered crop wages.

In-Season Farm Labor Reports (ES-223)

The Department of Labor's In-Season Farm Labor Reports estimate the number of local and migratory workers employed during the week which includes the 15th of the month, although some states apparently use different survey weeks. The Employment and Training Administration (ETA) requires that local Employment Service offices in areas with 500 or more seasonal workers or any temporary alien H-2A workers file monthly reports with ETA. ETA uses this data to operate its interstate job clearance system so, for example, if West Virginia apple growers request temporary foreign workers, ETA can determine whether unemployed American apple pickers are available in other areas. There is apparently no sanction on states for not making such estimates; 13 states reported 0 migrants in 1982 (or did not file reports), 19 states reported 0 in 1985, and 15 states including Washington reported no migrants in 1987.

The ES-223 reports divide farmworkers into three groups: local seasonal workers (employed at least 25 days in agriculture and deriving at least half of their total incomes from farmwork but not employed more than 150 days on any single farm); migrant workers (the subset of seasonal workers unable to return to their permanent residence at the end of the farm workday); and temporary H-2A workers. Agriculture is defined to include all crops (01); all livestock (02) except animal specialties (027); all agricultural services except veterinary services (074), animal specialty services (0752) such as farms boarding horses or dog kennels, farm labor contractors (0761), and landscape and horticultural services (078); it is not clear why the employees of labor contractors are not considered ES-223 farmworkers. Full-time students working or travelling with their (farmworker) families are counted as seasonal workers, but not full-time students travelling in organized groups.

The ES-223 data define migrants as the subset of seasonal workers "whose farmwork experience during the preceding 12 months" required an overnight stay away from home. ES-223 also defines migrant food processing worker as persons who were away from home overnight to work for up to 150 days with a food processing employer (SIC 201, 2033, 2035, or 2037) who packs, cans or freezes food products, and who earn at least half of their annual income in these three SICs.

State ES-223 reports list the crop activity, the stage of the work, and wages paid e.g., a typical report in California (ED&R Report 881-A) indicates that during the week ending September 12, 1985 there were 22,800 workers employed to harvest 166,000 acres of raisin grapes in Fresno county; that the raisin harvest was 75 percent complete; and that wages averaged $3.35 to $3.50 per hour and/or 13¢ to 16¢ per tray. ETA data are reported by Agricultural Reporting Areas (ARA's), not necessarily counties, and ARAs are the approximately 219 (multi) county areas throughout the U.S. which employ 500 or more seasonal workers or H-2A workers.

ES-223 data are regularly criticized by migrant researchers and then relied upon because they are the only data available which report migrant employment by state and month. The criticisms of ES-223 data are many, but all relate to the fact that ES-223 data are not generated by a statistical survey and thus contain unknown errors. For example, the ARA's do not cover all agricultural areas in the United States, so ES-223 data would be incomplete even if each state filed reports. More serious is the absence of a formal survey or estimating procedure--ES personnel with multiple duties have said that they simply reported last year's number or estimated migrants by counting out-of-state license plates in rural areas or reading about farmworkers in agricultural publications.

Three vignettes from California, Florida, and Texas give an idea of how the ES-223 data is collected. California requests monthly data from the 31 Agribusiness Representatives (ABRs) who are responsible for 42 of the state's 58 counties; however, several representatives fail to report each month. Local representatives have a handbook which details a procedure for estimating the number of workers based on the acres of each crop and estimates made in the 1960s of the hours of seasonal and other worker time needed to handle each acre. No statisticians are available to assist the ABRs, and the coordinator of the report acknowledges that its reliability varies from ABR to ABR.

Florida's Agricultural Service Representatives perform a similar function with similarly uneven results. Texas followed a similar procedure of compiling of 50 local reports until 1986, when budget cuts caused the Texas Employment Commission (TEC) to stop making ES-223 estimates. When DOL insisted on ES-223 data, the TEC treated the whole state as one Agricultural Reporting Area and assigned one person to the job of satisfying DOL by e.g., adjusting the previous year's data on the basis of weather reports and farm labor and production articles.

The reliability of what is reported to the ES-223 is uncertain. However, many states do not file ES-223 reports, so "ES-223" migrants are concentrated in a few states (Table 2.12). Between 1978 and 1982 California's share of migrants has been stable at about one-third of U.S. migrant man-months (one migrant employed during one fo the survey weeks), but, Michigan and North Carolina increased their shares of a shrinking number of U.S. migrant man-months, in part because states such as Washington quit reporting. Some of the 1978 to 1987 changes are dubious; it is doubtful that migrant man-months in Michigan almost doubled over these years, or that in Oregon migrant man-months were halved.

In 1987, the ES-223 reporting system was changed to report only "total migrants" although total migrants still appear to mean one migrant employed during one of the monthly reports. In 1987, there were 766,000 total migrants, and they were 32 percent intrastate and 68 percent interstate; California had 37 percent of the reported total migrants, Florida had 14 percent, and Texas had 1 percent.

The growing number of states with substantial agricultural production but no migrant or seasonal farmworkers makes it very difficult for researchers to continue to use the ES-223 data as a base which can be adjusted to arrive at the "true" number of migrants. When states such as Washington report no migrant and seasonal farmworkers in the ES-223 reports, how can researchers adjust the data to generate the "true" number of migrants? Making such adjustments requires another presumably better data base, and if such an alternative data base is available, there is no need to use ES-223 data at all.

Table 2.12

In-Season Farm Labor Migrant or ES-223 Data: 1978, 1982, 1987

ES-223 Total Migrant Man-Months

State	1978	Per Dist
California	307,100	33%
Florida	115,800	12%
Washington	69,900	7%
North Carolina	60,300	6%
Oregon	50,300	5%
Michigan	42,600	5%
Arizona	28,200	3%
Ohio	27,000	3%
New York	26,500	3%
New Mexico	19,700	2%
Top Ten States	747,400	80%
United States	934,400	100%

ES-223 Total Migrant Man-Months

State	1982	Per Dist
California	271,100	33%
Florida	152,900	18%
North Carolina	93,000	11%
Michigan	60,700	7%
Washington	59,600	7%
New York	22,100	3%
Arizona	18,100	2%
Idaho	18,000	2%
New Jersey	17,200	2%
Oregon	14,200	2%
Top Ten States	726,900	87%
United States	830,950	100%

ES-223 Total Total Migrants

State	1987	Per Dist
California	279,779	37%
Florida	109,213	14%
Michigan	83,166	11%
North Carolina	78,853	10%
Oregon	27,398	4%
Arizona	24,496	3%
New York	20,230	3%
New Jersey	18,890	2%
South Carolina	17,775	2%
Idaho	17,348	2%
Top Ten States	677,148	88%
United States	765,834	100%

Sources: ES-223 In-Season Farm Labor Reports, USDOL, Employment and Training Administration

A migrant man-month is one migrant employed during the survey week; this data cannot distinguish between a migrant employed 4 weeks during the month and a migrant employed only during the survey week

These estimates are made for multi-county Ag Reporting Units which employ 500 or more MSFW's or any H-2A workers Migrant workers do farmwork for 25 to 150 days annually, derive at least half of their annual earnings from farmwork, and are unable to return home at the end of a farm workday

In 1987, only data on "total domestic migrants" are reported; of these, 242,882 were intrastate, 522,952 were interstate, and 8914 were contract Puerto Rican. An additional 52,227 were "foreign" or presumably H-2, suggesting that this 1987 data is also man-months

MIGRANT EDUCATION

Most farmworker assistance programs acknowledge that their resources are not adequate to serve all of the workers eligible for assistance, and so they adjust establishment, household, or administrative data to distribute their federal funds across states and counties in proportion to each area's share of U.S. farmworkers. The Migrant Education is unique among assistance programs in its reliance on an in-house data source: instead of adjusting available farm labor data, the Migrant Student Record Transfer System (MSRTS) generates the data to allocate funds. This means that Migrant Education assistance funds reflect the efforts of school district recruiters as well as the employment of migrant families, that is, there can be a great deal of migrant activity but few children in the MSRTS if most of the migrants are solo men or if local recruiters do not enroll the children of migrant families.

MSRTS is a national clearinghouse for information on migrant farmworker children which was begun in 1969. The Federal Migrant Education Program established by the Elementary and Secondary Education Act in 1966 makes funds available to state educational agencies "to establish and improve programs to meet the special educational needs of migratory agricultural workers"(the children of migratory fishers were added later). State education agencies can use the MSRTS to transfer the records of migratory children from one school district to another in order to provide continuity for each child's educational program, and MSRTS records are used to allocate Migrant Education funds to states. Between 1982 and 1986, migrant education appropriations were $253 to $265 million annually.

Migrant education programs are justified by "the particularly difficult educational problems migrant children face" (Committee on Education and Labor, 1987, p. 35). The Congressional justification for migrant education programs rests on the alleged "extremely high" drop out rate of migrant children (90 percent in 1974, down to 50 to 60 percent in 1986); the typical 6 to 18 month lag of migrant children behind their expected grade levels; and the poverty, mobility, and non-English backgrounds of many migrant parents (Ibid., p. 35).

The Migrant Education Program serves the children of current and former "migratory agricultural workers." A migratory agricultural worker is a person who has moved within the past 12 months from one school district to another to enable him or her to obtain temporary or seasonal employment in an agricultural or fishing activity. Agricultural activity means an activity directly related to the production or processing of crops, poultry, or livestock for initial commercial sale or as a principal means of personal subsistence or activities related to cultivating or harvesting trees or activities directly related to fish farms; fishing activities are activities related to the catching or processing of fish or shellfish for initial commercial sale or subsistence. The children of migratory processing workers (who are often employed in industries with nonfarm SIC codes) are not distinguished in the MSRTS data. The MSRTS data theoretically count migrant children (continuously) where their parents work or where they live, since the movements of children (parents) are recorded in the MSRTS; the HFWF, in contrast, counts migrant workers once each year, presumably where they live. However, the MSRTS treatment of withdrawal days, when the student is not in an MSRTS-linked classroom because e.g. he is in a public or private school but not in a Migrant Education program, he is in the U.S. but not in school, or is out of the U.S., complicates the interpretation of whether MSRTS actually "tracks" migrant children.

A "currently migratory child" is a child whose parent, guardian, or another member of the family is a migratory agricultural worker or fisher or a child who himself or herself has satisfied the definition of migratory worker within the past 12 months. A "formerly migratory child" is a child who was eligible to be counted as a migratory child within the past five years, lives in an area served by a Migrant Education program (this residence requirement has been deleted recently), and has parents who approve having the child be considered a migratory child.

Many of the ambiguities in these definitions are clarified in various policy statements. For example, "to move within the past 12 months" means to establish a residence in another school district and normally requires at least a 48 hour stay, i.e., commuting across school district lines does not constitute a move. The move must be made to enable the parent to find temporary or seasonal farm employment: if the parent does not find farm employment, the children may still qualify for Migrant Education programs if the purpose of the move was to find farm employment. Temporary and seasonal farmwork mean "not permanent," or farm jobs which do not continue indefinitely.

Until 1987-88, the primary focus of Migrant Education programs was not current and former migratory children 5 through

17 who were in school. States could also serve pre-school and 17 to 21 year-olds, but the funding base was 5 to 17 year olds.

Allocations to states for 1988-89 will be based on the number of current and former migratory children in the MSRTS aged 3 to 21 in calendar year 1988. This new funding formula has changed the recruitment incentives of State and local school personnel. The 1988-89 priorities of migrant education are: (1) school-aged migratory children; (2) formerly migratory school-aged children; (3) pre-school migratory children; and (4) formerly migratory pre-school children. In 1989-90, all currently migratory children will have priority over formerly migratory children.

MSRTS data are available for 49 states and several U.S.-administered territories. All states except Mississippi have mandatory school attendance laws, but Migrant Education programs typically employ outreach workers to recruit migrant students instead of waiting for state education officials to find migrant children and compel them to attend school. States vary in their estimated proportions of eligible students served; in 1985, Migrant Education reported serving 382,253 full-time equivalent migratory children who were 5 to 17, of an estimated 530,367 eligible children under age 21, or 72 percent. Migrant Education funds are allocated to states on the basis of their shares of full-time equivalent students credited to each state in the MSRTS.

The MSRTS divides migratory children into six groups based on where their parents worked (agriculture or fisheries); the type of migration (interstate or intrastate); and the child's current migratory status (current or former as determined by the person in the family who does or did farmwork). Data are reported by state for full-time equivalent students--a full-time equivalent student is 365 days of enrollment in MSRTS.

Migratory students are entered into the MSRTS after they are enrolled by a local school district, and migratory students remain credited to that school district until they are identified in another school district or state. This means that once enrolled, a migratory student remains "enrolled" in that school district for up to 13 months, even if the student "withdraws". The withdrawal days of migratory students are added to the days that migratory students actually attended school to determine total migrant FTE; withdrawal days are included in the count because Migrant Education funds are allocated on the basis of the number of days that migrant children reside in a state; not just the number of days that migrant children attend school in a state. This inclusion of withdrawal days has prompted some grumbling about states that are stopping off points for migrants (such as Arkansas) enrolling migrant students and getting credit for them for up to 13 months unless the migrant children re-enroll in an upstream state such as Michigan or Ohio.

Migrant Education (ME) funds are concentrated in Texas and California; these states received over half of ME funds in 1982 (Table 2.13). California had the most students whose parents are current migrants, but Texas had about 20,000 more formerly-migratory students in 1982. The top ten states obtained about three-quarters of all ME funds, and the top ten ME states include several states not usually associated with migrant farmworkers, such as Louisiana and Massachusetts; these states include relatively large numbers of migrant fishers.

MSRTS is a system which provides detailed data on the children of current and former migrants, and thus should be able to generate indirect data such as family size and migration patterns. However, the migrant characteristics in the MSRTS reflect the efforts of recruiters to enroll migrant children; if recruiters systematically enroll only children from families with certain characteristics, then the MSRTS will reflect the characteristics of only that part of the migrant population. More important, MSRTS data most likely reflect the characteristics of migrant families, not solo male or skilled migrants. Thus, if particular groups of migrants are more likely to migrate as families (Texas Hispanic families versus single Florida Blacks), then MSRTS migrant data give distorted picture of the overall migrant workforce.

Migrant Education also funds a High School Equivalency Program (HEP) and a College Assistance Migrant Program (CAMP). Both HEP and CAMP make grants to institutions of higher learning or agencies that cooperate with such institutions to help the children of migratory farmworkers. HEP funded 20 projects that served about 2,700 students in 1986-87 at a cost of $6 million or $2,233 each; CAMP funded five projects that served 370 students at a cost of $1.1 million or $3,103.

The HEP and CAMP programs make grants to applicants who recruit, counsel, and place migrant children (presumably persons 17 or 18 and older and thus beyond the age of compulsory school attendance). HEP and CAMP may provide participants with room and board and stipends to cover their personal expenses.

The 1987 School Improvement Act (H.R.5) adopted several new programs for migrant children. Chapter 1 programs (previously Title I programs of the Elementary and Secondary

Table 2.13

Migrant Education Enrollment and Funding: 1982

State	Migrant Students 5 to 17	Percent Interstate	FTE Students 5 to 17	Percent Interstate	Former Mig Students 5 to 17	Former FTE Mig Students 5 to 17	Migrant Educa Funds -$000-	Per Dist
Texas	66,231	58%	48,047	53%	74,302	68,111	63,133	26%
California	75,523	53%	61,532	52%	54,038	49,059	62,281	25%
Florida	26,467	83%	19,057	80%	15,872	14,959	17,907	7%
Washington	9,627	68%	6,507	61%	6,666	6,094	9,350	4%
North Carolina	5,460	61%	3,029	57%	9,083	6,794	8,336	3%
Arizona	7,983	72%	5,432	68%	7,229	6,653	6,561	3%
Lousiana	2,898	50%	1,703	46%	6,553	6,532	6,111	2%
Oregon	5,127	71%	3,413	64%	3,978	3,557	5,488	2%
Michigan	9,225	88%	4,558	80%	2,296	2,039	5,477	2%
Massachusetts	1,194	89%	871	89%	5,158	4,902	5,441	2%
Top Ten States	209,735		154,148		185,175	168,700	190,084	77%
United States	277,812	65%	190,129	59%	238,366	216,312	245,653	100%

Source: Office of Education, MSRTS Enrollment Summary

U. S. total excludes Puerto Rico, D.C., and other federal territories; there is no Migrant Education program in Hawaii

FTE is days of migrant student enrollment divided by 365

Migrant children had parents who moved across school district lines for temporary or seasonal agricultural or fishery job

Former migrant students had parents who were migrants sometime within the past 5 years

Education Act of 1965) are designed to serve "low-achieving students in poor areas;" and these programs allocated almost $4 billion to states in 1987-88, and served almost 5 million children in 1984-85. H.R.5 proposed a $50 million Even Start program for disadvantaged pre-school children, and Section 1052 reserved 3 percent or $1.5 million for migrant children.

H.R.5 also proposed a $2 million 12 member three-year Commission on Migrant Education to examine "the changing demographics of the migrant student population in an effort to assure that patterns of migrancy are anticipated and the children are served to the best extent possible" (Committee on Education and Labor, 1987, p. 38). This Commission is also charged with evaluating the need for a National Center for Migrant Affairs, and to help develop a blueprint of changes needed in migrant education programs in the 1990s.

SUMMARY

There are three critical definitions involved in enumerating migrant farmworkers: the definition of a farm, of a farmworker, and then a migrant farmworker. Farms are usually considered places which sell farm products worth $1,000 or more annually, and farmworkers are often all persons who are employed by farm operators, whether they are field workers, equipment operators, or office clerks. Migrants are usually defined as the subset of farmworkers who cross a geographic boundary such as a county or state line and stay away from home overnight to do farmwork for wages. Some migrant definition expand farmworker and thus migrant to include persons employed by agricultural service firms which employ workers to work on farms, such as a labor contractor who hires peach pickers, and some extend migrant status to packing shed and food processing workers.

Data on farmworkers are obtained from establishments or employers, households or workers, and from the administrative records of government agencies or assistance programs. The major establishment data sources include the Census of Agriculture, the Quarterly Agricultural Labor Survey, and the Farm Costs and Returns Survey; COA and FCRS obtain data primarily on farmers' expenditures for labor, while QALS obtains four weekly snapshots of farm employment and hourly earnings.

Household data are collected in the Census of Population and the Current Population Survey. Both of these household surveys, which are used to chart trends in the nonfarm labor force, are inadequate sources of data on especially migrant and seasonal farmworkers. The COP bases its employment information on a March reference week, when there are relatively few farmworkers employed, and the CPS sample includes relatively few farmworker households.

Administrative data are obtained from government agencies and assistance programs. Major administrative data sources include the data that large farm employers report to Unemployment Insurance authorities (ES-202) and the data reported by local Employment Service offices to the Department of Labor (ES-223); however, both of these administrative sources include only a partial picture of the farm labor market. The Migrant Student Record Transfer System is the most sophisticated and expensive source of migrant farmworker data, but it concentrates on the children of migratory farmworkers and probably reflects only the distribution of migrant families with children.

The state-by-state distributions generated by these establishments, household, and administrative data sources for 1982 are compared in Table 2.14. Farmworker activity is concentrated in a few states in all of the distributions; the top 10 states include 53 to 87 percent of U.S. migrant activity. California was ranked first in 1982 in every distribution except Migrant Education, and California's share of U.S. farmworker activity ranges from a low of 17 percent in the 1980 COP enumeration of farmworkers to a high of 46 percent of fruit and vegetable labor expenditures. Thus, California appears to include 20 to 50 percent of all farmworker activity.

California dominates most farmworker rankings, but the ranking of the other top ten states changes considerably in different data sources. The COA- and FCRS-generated labor expenditure data such as BEA make Florida, Texas, and Washington major farm labor states, but they also include midwestern states such as Wisconsin, Illinois, and Iowa among the top ten. The COA seasonal crop worker tabulation puts Appalachian states such as Kentucky, North Carolina, and Tennessee among the top ten, while COA fruit and vegetable labor expenditures include Pennsylvania, New Jersey, Minnesota, and New York as major farm labor states.

UI-covered employment and ES-223 migrants are very concentrated, but both of these administrative data sources are

Table 2.14
Farm Labor Distribution Indicators for the Top Ten States

Census of Agriculture — Tot Crop Labor Expendits

State	1982
California	33%
Texas	5%
Florida	11%
Washington	5%
Wisconsin	1%
New York	2%
North Caroli	3%
Illinois	2%
Pennsylvania	1%
Iowa	2%
Top Ten Stat	64%
U.S. Total	100%

Census of Agriculture — Crop Seasonal Workers

State	1982
California	22%
Kentucky	7%
North Caroli	8%
Washington	8%
Texas	3%
Iowa	3%
Tennessee	3%
Florida	4%
Minnesota	2%
Oregon	4%
Top Ten Stat	64%
U.S. Total	100%

Census of Agriculture — FVH Tot Labor Expenditure

State	1982
California	46%
Florida	14%
Washington	5%
Pennsylvania	3%
Texas	3%
New York	3%
Michigan	3%
Oregon	3%
Arizona	2%
New Jersey	2%
Top Ten Stat	83%
United States	100%

Census of Agriculture — Farm Labor Expense — BEA

State	1982
California	20%
Texas	6%
Florida	6%
Washington	4%
North Caroli	3%
Wisconsin	3%
Illinois	3%
Iowa	3%
New York	3%
Pennsylvania	2%
Top Ten Stat	53%
United States	100%

UI-covered Employ

State	1982
California	44%
Florida	14%
Washington	4%
Texas	3%
Pennsylvania	3%
New York	3%
Oregon	2%
Illinois	2%
New Jersey	2%
Michigan	2%
Top Ten Stat	80%
United States	100%

ES-223 Migrant Man-Months

State	1982
California	33%
Florida	18%
North Caroli	11%
Michigan	7%
Washington	7%
New York	3%
Arizona	2%
Idaho	2%
New Jersey	2%
Oregon	2%
Top Ten Stat	87%
United States	100%

Migrant Education Funds

State	1982
Texas	26%
California	25%
Florida	7%
Washington	4%
North Caroli	3%
Arizona	3%
Lousiana	2%
Oregon	2%
Michigan	2%
Massachuset	2%
Top Ten Stat	77%
United States	100%

Census of Population — Farmworkers

State	1980
California	17%
Texas	7%
Florida	6%
Wisconsin	4%
Iowa	3%
North Caroli	3%
New York	3%
Minnesota	3%
Illinois	3%
Washington	3%
Top Ten Stat	53%
United States	100%

influenced by state decisions to extend coverage to workers on smaller farms (UI) or to count migrants (ES-223). The last two sources, ME and the COP, reflect other factors: ME reflects both the success of recruiters as well as the presence of migrant families and their children in a state, and the COP reflects the employment of farmworkers in March, hence the inclusion of Wisconsin, Iowa, New York, Minnesota and Illinois.

This comparison of farmworker distributions across states emphasizes the truism that no two data sources agree on the distribution of farmworker activity across states. As the review of data sources demonstrate, one reason for these conflicting distributions is that each data source is estimating a different slice of farmworker activity. Thus, researchers must be careful to justify the source(s) of farmworker data that they select. Chapter 3 reviews studies based primarily on ES-223 data, and Chapter 4 presents a migrant farmworker count and distribution based on UI and labor expenditure data.

NOTES

1. This chapter is based on several reports which have assessed farm labor data, including Daberkow (1986), Tweeten (1979), and Holt (1987).

2. There is a problem with COA county-level data in states such as California, which had 1,000 farms with more than one establishment or unit in the 1969 COA. The COA simply tabulates each sub-unit in the county in which it is located, thus understating the degree of concentration in agriculture. Thus, Newhall Land and Farming, an NYSE-listed company with operations in 5 California counties, appears in the COA as 5 operations, not the one farming company that it is.

The Census of Manufactures does not confuse companies and establishments because it reports companies and establishments separately. The fact that there are today more than 1,000 California farming companies that have more than one establishment or sub-unit emphasizes just how much California farms are factories-in-the-fields. I am indebted to Don Villarejo for clarifying county-level COA procedures.

Chapter 3

Bottom-up Migrant Studies

The federal government developed programs to assist migrant farmworkers and their families in the mid-1960s, and since then there have been a series of studies which distributed migrant workers and their dependents across states in order to allocate the approximately $500 million annually in federal migrant assistance monies. These studies have left a sorry legacy: the typical study of the number and distribution of migrant farmworkers begins with a survey of farm labor data sources, finds all of them deficient, and then because of limited funds and time begins to "adjust" one or several of the data sources in order to produce a count and distribution of migrants.

There is no scientific or even generally agreed upon methodology to adjust available farm labor data, and migrant studies reflect a variety of often contradictory estimation and distribution procedures. Some studies adjust only one data source and some adjust several; some "correct" the data for particular states on the basis of local observations or case studies; and some studies ignore farm production indicators while others rely on them. As might be expected, no two studies agree on the adjustment methodology or the number and distribution of migrant farmworkers.

Most studies are based on a similar methodology, and an example will illustrate the usual adjustment procedure. Most studies employ a "bottom-up" estimation procedure: they begin with an estimate of peak migrant employment in each state, usually based at least in the part in DOL ES-223 data, adjust it, add dependents, and then estimate each state's peak migrant population.

    A hypothetical example illustrates the typical adjustments.
Assume that the ES-223 estimates that a peak 50,000 migrants are
employed in September in California. The first step is to increase
this peak migrant employment estimate by a turnover factor: if
50,000 migrants are employed during the survey week in
September, turnover implies that a higher number of persons
actually filled the estimated 50,000 migrant jobs estimated by the
survey. A typical turnover ratio is 1.25, meaning that for every four
jobs counted during the survey week, five persons were hired. This
is a very high rate of worker turnover: a business which offered
four year-round jobs but had to hire one new worker each week
would hire a total of 56 individuals during the year, or 14 persons
per year-round job.[1] A 1.25 weekly turnover ratio raises the 50,000
peak migrant estimate to 62,500 survey-week workers.
    These 62,500 workers are then reduced to migrant households
by assuming that each migrant household supplies 1.2 to 2.0
migrant workers. If each household supplies 1.5 workers, then
62,500 workers becomes 41,667 households. These households
are separated into family and single person units, and a typical (and
critical) division is to assume that 80 percent of all migrant
households are families and 20 percent are single persons, or
0.8 x 41,667 = 33,334 family households.
    The penultimate step is to calculate the number of dependents in
migrant family households. Most studies assume that each migrant
family has two to seven dependents, and if we assume five for
illustrative purposes, then 33,334 x 5 = 166,668 dependents in
migrant households. Finally, 62,500 peak migrant workers plus
166,668 dependents can be summed to obtain a total 229,168
migrant workers and dependents.
    There are several variations on this bottom-up estimation
procedure. The initial peak month estimate can be derived from the
ES-223 data or from assistance program administrative data. The
turnover factor can be applied to increase peak employment and then
a duplication factor included to reduce the migrant count because
mobile workers may be counted on two farms during the survey
week. However, almost all of the studies reviewed divide workers
into the 80 percent family--20 percent single categories, and then
multiply family migrants by dependent factors of 4 to 6. If
80 percent of the households are families which each have five
dependents, and each household supplies 1.5 migrant workers, then

100 migrant workers generate a total migrant population of 367, and dependents are three-fourth of this migrant population.[2]

This "bottom-up" estimation procedure yields an estimate of the peak number of migrants in each state. Such estimates are of dubious reliability because of deficiencies in the ES-223 data (see Chapter 2) and may be of limited use for allocating assistance funds because such peak month employment estimates do not indicate the duration of migrant employment in each state, do not separate migrant workers from dependents, and may provide a misleading aggregate migrant profile.

Peak month employment estimates are based on migrants at work and turnover during a one-week snapshot of the labor market. Such a snapshot procedure cannot distinguish between an area with a short but intense harvest season such as Oregon strawberries and a longer harvest such as California citrus with few peaks and valleys. The implicit implication of allocating assistance funds on the basis of employment and turnover during one week is that migrants employed in areas with peak employment during the survey week need the most assistance.

This bottom-up adjustment procedure favors areas with peak employment during the survey week; it also favors areas with large migrant families employed during the survey week. Most migrant studies assume that migrants are typically large families, so that 60 to 80 percent of the migrant population in an area are dependents, not workers. In the example above, if family migrants were half rather than 80 percent of the migrant workers, the migrant population falls 27 percent from 367 to 267, even though the peak week workforce remains unchanged at 100. Similarly, if family migrants remain 80 percent of peak week migrant workers, but they average only two instead of five dependents per family, the migrant population falls 36 percent. These calculations highlight the importance of the migrant family assumptions in migrant population counts.

The third issue raised by the usual bottom-up estimation procedure is that it may generate a misleading aggregate migrant profile. The 100 peak workers who generate a migrant population of 367 in the example above may work in a succession of states, so that Florida, North Carolina, Delaware, and New York are each credited with 367 migrants, or 100 workers generate a total migrant population of 367 x 4 or 1,468 in four states, so that patterns of migration and state borders may unduly influence the migrant population estimate in these bottom-up studies.

No migrant study is perfectly accurate because most data sources do not enumerate the migrant population of interest. However, this overview suggests that migrant studies should separate workers from dependents, justify their family assumptions, and perform sensitivity analyses on how peak week employment and family assumptions affect the migrant profile they paint. As this chapter demonstrates, few studies adopt these clarifying and sensitivity procedures.

MIGRANT STUDIES COMMISSIONED BY THE MIGRANT HEALTH PROGRAM

The 1962 Public Health Service Act authorized funds to establish clinics and to train medical staff to improve the health services available to domestic agricultural migratory workers and their families. In 1970, the Migrant Health Program was expanded to include "other seasonal farmworkers and their families." Today, the Migrant Health Services Program spends about $50 million annually to serve migrant and seasonal workers and their families through about 130 clinics.

The Migrant Health Services Program serves "migratory agricultural workers, seasonal agricultural workers, and the members of the families of such migratory and seasonal workers." A migratory agricultural worker is "an individual whose principal employment is in agriculture on a seasonal basis, who has been so employed within the last 24 months, and who established for the purpose of such employment a temporary abode;" a seasonal worker also has his principal employment in agriculture on a seasonal basis but does not established a temporary abode. Agriculture is defined as crop farming and the agricultural services related to crop farming that are performed on a farm, i.e. livestock workers are not eligible for Migrant Health Program services.

The legislation authorizing the Migrant Health Program was modified in 1975 to emphasize primary health care in "high impact" migrant areas--defined as areas with 4,000 or more migrant and seasonal workers and their dependents for more than two months in any calendar year. It appears that a "high impact area" may be a single county or multi-county region, but studies for Migrant Health usually produce county-level estimates of migrants and their dependents and seasonal workers. There are approximately 3,150 U.S. counties, and Migrant Health considers about 900 to be high impact migrant counties.

The studies of migrant and seasonal farmworker populations apparently play only a limited role in allocating Migrant Health funds to farmworker clinics. Clinics must justify their requests for funds by estimating the number of migrant workers and dependents to be served, but it appears that applicants for funds are free to use estimates from the studies commissioned by Migrant Health, other migrant farmworker estimates, or to simply assert that their clinic records are the best source of local data on migrants when estimating the population to be served and justifying their requests for funds.

The migrant studies commissioned by the Migrant Health Program may have had more influence outside of the migrant health program than inside it. The most significant migrant count and distribution report was published in 1973, since it was the benchmark for subsequent mid-1970s reports which continue to determine the number and distribution of migrants for several assistance programs in the mid-1980s.

The 1973 Migrant Health Program Target Population Estimates

The 1973 report was prepared by Migrant Health staff and consultants. The basic data sources were local estimates and ES-223 migrant data. Peak month ES-223 job estimates were adjusted bottom-up reflect workers and dependents, distributed to counties, and then simply added together to generate state and national estimates. The 1973 report estimated a peak 1.1 million migrants and dependents. To correct for multiple job holding and turnover (migrants appearing in two survey areas during the year), the 1.1 million U.S. migrant population estimate was reduced to 700,000, or 653,032 excluding Puerto Rico. Eight states, including Georgia, Mississippi, and Tennessee, were reported to have no migrants or dependents. Even though the 1973 report was constructed bottom-up from migrant worker estimates, only the total population of migrants and dependents was reported for each state.

The 1973 report relied primarily on local estimates of peak migrant numbers in each county, although these local estimates were based in part on ES-223 data. When ES-223 data were utilized, factors were developed to determine the number of workers and dependents per migrant household. These factors were derived from a variety of sources: the states of Iowa, Kansas, Missouri, and Nebraska were assumed to have 3.5 migrant workers and 1.8 dependents per household based on "composite data" from health projects in these states. The Texas ratio of 3.4 workers and 4.1 dependents per migrant household was obtained from the Texas Governor's Office and applied throughout the midwest from Illinois, Wisconsin, and Michigan to Oklahoma and New Mexico.

The average migrant worker and dependent ratios per household deserve scrutiny because they influence the estimated migrant population and they were used in several subsequent studies. Three states had unique factors: New York (three workers and five dependents per household from HEW records); Utah (1.7 and 4.5 from Migrant Council); and California (2.5 and 5.3 from Migrant Housing Center reports). There were six household and dependent factors developed and applied to more than one state:

These household and dependent factors prompt two observations: first, none are derived from a scientific study or sample and second, all of the sources listed are likely to inflate average migrant worker- and dependent-per-household factors. For example, worker and dependent factors derived from records of public labor camps in California should indicate higher -than-average dependent factors because large families have the most incentive to search out the subsidized housing and educational, health, and social services available in the camps. It is hard to believe that the average size of migrant households in Texas (and thus throughout the Midwest) is almost 11; even the almost nine persons per household factor in the Plains and Mountain states strains credulity, especially when it is remembered that these are averages, so that for every migrant household with one migrant worker and two dependents there must be another with six migrant workers and 13 dependents.

The peak migrant estimates for each county were summed to obtain state totals, and these peak state totals were adjusted downward to reflect intrastate migrants. These intrastate mobility ratios apparently varied, but the sum of peak migrant worker employment across all states (215,200) was reduced to 148,100 individuals employed during each state's peak month, suggesting that migrants averaged 1.45 interstate employers. Peak month migrant worker employment, apparently from ES-223 data, ranged from 50,100 in California in October 1973 to 15,300 in Texas in July to 14,000 in Florida in March.

State peak migrant estimates were summed to obtain regional and national totals. Regional and national migrant totals were not adjusted for the duplication that would occur if the same individual were included in the peak estimate of several states because the researchers believed that their basic county and state estimates were undercounts.

The 1973 report includes an appraisal of the various farm labor data sources and a short bibliography. The 1973 report is distinguished by the apparent participation of persons responsible for farm labor data in USDA and DOL. The 1973 report generated the fewest migrant workers (an estimated 148,100 employed during each state's peak month) and the fewest total migrants and dependents in the 50 states (653,032) of all of the migrant studies reviewed here.

The 1978 Migrant Health Target Population Estimates

The 1978 Migrant Health report was prepared by InterAmerica Research Associates and published in 1980. State and county estimates are based on ES-223 data, corrected for "variances" and then adjusted to include dependents, unemployed workers, and nonfarm processing workers (pp. 1-2). The report estimated that there were 800,000 migrants and dependents in 910 U.S. counties in 1978, an increase of 15 percent from 1973. An additional 1.9 million seasonal workers and dependents were estimated to be in these migrant-impacted counties, for a total 2.7 million migrant and seasonal workers and dependents.

The 1978 report first attempted to determine the accuracy of ES-223 data for each county or area by contacting ES representatives and local migrant agencies. If an undercount was suspected, the ES-223 estimate was increased by 20 percent (p. 23) to account for unemployment on the 15th of the month or an employment peak that was not on the 15th. In some states, another (unspecified) data source was used to correct ES-223 data or to adjust it.

The major adjustment which affects migrant population estimates are the household and dependent factors. The 1978 report used the same migrant worker-and dependent-household factors that were developed in the 1973 Migrant Health report, but InterAmerica apparently assumed that all migrant workers are in family units.

Table 3.1

Migrant Worker and Dependent Factors Used in the 1973 Migrant Health Report

Region	Migrant Workers per Household	Dependents per Household	Source
New England (8 states)	1.0	1.5	MSRTS Data
Mid-Atlantic (5 states)	1.6	4.0	Delmarva Ecumenical Agency
Southeast (8 states)	1.6	4.3	Migrant and Rural Farm Labor Housing Study: Florida, 1974
Midwest (14 states)	3.4	7.5	Texas Governors Office
Plains (5 states)	3.5	5.3	Composite Migrant Health Data
Mountain (4 states)	2.4	5.7	Colorado Migrant Council

Source: Migrant Health Program Target Population Estimates, 1973.

Table 3.2

Migrant Health and Lillisand Estimates of Migrant Workers and
Dependents for Selected States*

	Migrant Health (1973)	Lillisand (1976)
Arizona	4,613	17,714
Arkansas	5,274	6,066
Connecticut	5,179	6,031
Delaware	5,437	9,379
Florida	76,345	166,964
Idaho	14,462	25,134
Illinois	24,247	41,826
Iowa	1,411	2,435
Kansas	4,593	8,924
Louisiana	8,984	10,332
Maryland	4,593	7,871
Massachusetts	2,884	3,677
Michigan	51,776	77,664
Minnesota	25,193	43,457
Missouri	1,187	2,048
Nebraska	3,234	5,579
New Jersey	11,146	19,227
Oklahoma	7,853	13,550
Rhode Island	158	171
Texas	153,731	318,225
Total	412,375	786,274

*These are the states for which Lillisand's 1976 estimates are based on the 1973 Migrant
Health Target estimates.

The next adjustment was to reduce the number of migrants by 25 percent because of double-counting, which occurs if ES representatives include the same person in two or more counties or reporting areas during the survey week as the workers move from farm to farm, and to increase the number of seasonals by 1.25 to account for worker turnover. These adjustments were made for some but not all states (p. 24).

State totals were summed for the ten multi-state federal regions, and these regional totals were then adjusted to account for intra-regional migration. Regional migrant population estimates were reduced by 5 to 33 percent; e.g., the estimated total 6,340 migrants and dependents in Iowa, Kansas, Missouri, and Nebraska was reduced by 5 percent to 6,000 to account for migrants who were counted in two or more of these states. The adjusted "total regional estimate" of migrants and dependents (839,700) was reduced by 5 percent to account for inter-regional migration (for example, a person counted in Florida and to Mid-Atlantic states) to yield 800,000 "unique" migrants and dependents in the United States.

The 1978 report includes a preface by the Office of Migrant Health which advises that the report can be used primarily to determine the location of high-impact migrant areas. Migrant Health administrators warned that "these estimates are not sufficiently reliable to form the sole basis for determining" such areas (p. iii). Local programs are asked to use data sources such as ES representatives, migrant education, "local farmer and farmworker organizations," and other MSFW organizations to estimate the number of migrants and dependents (p. iv). These estimates are to be sent to the regional migrant health office to justify an application for funds.

The 1978 report implicitly acknowledges the limitations of the data by rounding numbers freely (e.g., reducing the adjusted 839,700 migrants and dependents by 5 percent yields 797,715, not 800,000). However, the main problem with the report is its apparently arbitrary adjustment factors, for example, increasing ES-223 migrant data by 20 percent to account for an undercount is based on "verification on other sources" (p. 22). Similarly, there is no real justification offered for subsequent reductions of 25 percent to account for intrastate migrants, a variable percent for intra-regional migration, and 5 percent for inter-regional migration.

Methodology for Designating High Impact Migrant and Seasonal Agricultural Areas (1985)

The 1985 Methodology report ignored the ES-223 data and attempted to estimate the number of migrant and seasonal workers in each county on the basis of crop and acreage data. The contractor (HCR) estimated the average hours needed to harvest certain crops, decided that migrants worked only in harvesting and did an average 35 percent of the harvest work, assumed that each migrant worked 576 hours per season, and then generated the number of migrants in each county.

The 1985 report includes two tables for each state. One table estimates the maximum number of all types of farmworkers and then seasonals, migrants, and migrants plus dependents for each county (no state totals are provided in the report). The second table for each state includes the major labor-intensive crops in each county.

The HCR report includes an 8-step "methodology" to determine the maximum number of farmworkers, seasonals, and migrants in each county. The methodology requests that local agency estimators:

1. Determine the county or area of interest
2. Determine the acreage of "labor-intensive" crops in each area or county (HCR recommends checking for 70 crops that range from apples and oranges to potatoes and coffee (sic!)-- the most labor-intensive crop).
3. Check with the local Extension Service to confirm which crops are labor-intensive.
4. Determine the hours necessary to harvest each acre (the "methodology" considers and ignores the hours needed "to plant, maintain, and process crops" by asserting on p. 27". .. 90 percent of migrant and seasonal farmworkers are used only to do harvesting.")
5. Determine the length of the harvest.
6. Calculate the number of workers needed to harvest the labor-intensive crops. A sample calculation is provided, which can be simplified as follows:

Strawberries: - 1 acre x 2,000 hours/acre = 2,000 hours.
- assume 8 hours/day/person x 72 days/season = 576 hours contributed by each worker.

- farmworkers needed = 2,000 hours ÷
576 hours per worker = 3.5 workers per
acre.
7. Divide this "maximum farmworker estimate" into migrants
and seasonals by assuming that 65 percent or 2.3 workers
are seasonal and 35 percent or 1.2 workers are migrants.
The actual percentage migrant was estimated for 42 states on
the basis of the 1978 Migrant Health report and 1980
Colorado Migrant Council estimates, and ranged from
2 percent of the farmworkers in Missouri who were
considered migrants to 95 percent migrants in Wisconsin.
Then the methodology increases the number of migrants by
17.1 percent to account for nonworking dependents.
8. Determine the maximum number of workers needed at any
one time in the area by summing up the workers needed for
each day of each crop harvest.

The HCR report concludes that this model "is theoretically sound,
gives constant results (sic), is easy to understand and use and makes
use of data already available" (p. 33).

The HCR report notes that this "methodology" is based on four
"reasonable and necessary" assumptions: crop acreages do not vary
significantly from year to year; the state-by-state migrant ratios are
reasonably accurate; crops requiring "less than five workers to
harvest," crops harvested by family members, and all activities
except harvesting can be excluded from the calculations, although it
is not clear that such exclusions were in fact made; and there are 17
dependents accompanying each 100 migrant workers. HCR
concludes that Migrant Health should work with "regional
consultants" in each area to help local clinics make migrant
farmworker estimates.

It it not clear how Migrant Health used the HCR report. Migrant
Health apparently allocated its funds in 1986 to regional groupings
of states on the basis of clinic service data (80 percent) and HCR
estimates (20 percent). The regional Migrant Health offices re-
allocated these funds to states. Local applicants for regional Migrant
Health funds must prepare a "need/demand assessment" that
includes the completion of six worksheets which detail, inter alia,
the population of migrant and seasonal farmworkers and dependents
in the area to be served. Worksheet 1 includes the farmworker
estimates, and applicants "are urged to use the easiest and least
costly approach" to estimate the number of migrant farmworkers in
the area, including ES-223 data, Migrant Education and Head Start

data, local outreach data, and/or the HCR methodology. The example provided to guide applicants includes calculations based on the HCR methodology, ES-223 data, and clinic enrollment data. The very different results for the sample county--2036, 900, and 3500--are discussed and the 3,500 clinic enrollment estimate is adopted because the applicant asserts that 3,500 migrant farmworkers and dependents are registered and that this is "98 percent of all migrant farmworkers" in the area (p. A-II-7).

The HCR study was apparently a two-year effort to produce a "practical and theoretically sound model" to estimate migrant farmworker populations, but the summary of the report above indicates that it is neither practical nor theoretically sound. It seems impractical to expect a health clinic applicant to perform the tedious calculations that would be necessary to determine acreage and harvest hours for the more than 30 vegetable crops grown around Salinas (California) even if such data were available. It is not theoretically sound to make the estimates depend on fallacious national assumptions such as: all harvest work is done by hired workers in "labor-intensive" crops; the harvest hours per acre of e.g., strawberries are 200 throughout the United States [total hours are estimated to be 2,000 to 3,000 hours per acre in the areas of California which grow almost three-fourth of U.S. strawberries]; migrants throughout the United States work eight hours daily and six days weekly; all migrants work the same number of hours in each crop and area; and there are 17 dependents for every 100 migrants. These flawed assumptions are compounded by arithmetic errors and the exclusion of labor-intensive nonharvest activities, for example, irrigation uses one-sixth of the man-hours required in California agriculture and many irrigators are migrants.

The HCR report does not include state or national totals for any of its county farmworker estimates, but tedious addition does yield a few numbers that should surprise migrant farmworkers. HCR estimated that the maximum number of migrant and seasonal farmworkers needed in 1982 was 1.7 million, and these workers were 63.6 percent migrant. However, migrancy averaged only 35 percent across states, illustrating the arithmetic errors throughout the report. For example, Arizona is reported to have 54 percent migrants, yet on the same page (32) the percentage migrant in Cochise County, Arizona is listed as 27 percent. It is impossible to reconcile the various HCR numbers presented even if one accepts the "methodology."

HCR found more migrants in Kentucky than in any other state: 293,278 migrant workers. California, by contrast, had a maximum 148,630 migrants; Washington 109,486; North Carolina 102,852; and Texas only 18,318. It is not clear how HCR derived these "maximum migrants needed" numbers from the data it presented. For example, Kentucky is listed as "needing" a maximum 325,863 farmworkers. HCR on page 32 says that 10 percent of Kentucky's farmworkers are migrants, but HCR actually reports that 293,000 or 90 percent are migrants. Apparently HCR reversed the 90 seasonal 10 migrant percentages (p. 32).

Other state percentages also appear to be reversed. California "needs" a maximum 198,163 farmworkers, of whom 25 percent are migrant (p. 32), but HCR actually reports 148,630 migrants or 75 percent of the total. Washington's maximum 144,000 farmworkers becomes 109,486 migrants, or 76 percent migrant when the table on page 32 says 24 percent migrant. It appears that HCR often interchanged migrant and seasonal when multiplying its maximum farmworkers by its state migrant percentage.

The HCR "methodology" is also flawed by its failure to cross-check the results with any other data source. One logical cross-check is the number of workers employed in the Census of Agriculture: each time a worker is hired by a farm employer, a "worker employed" is recorded. Since some workers are known to be hired by several farmers, the number of farmworkers in a state should not exceed the COA "workers employed" number. In Kentucky, HCR reported a maximum 325,863 farmworkers and 293,278 migrants; however, the Census of Agriculture reported that Kentucky's 52,462 farm employers employed a total of only 267,566 workers, including 247,808 for less than 150 days. Thus, HCR's "methodology" appears to be flawed for estimating workers and migrants in Kentucky, since it is improbable that there are more unique farmworkers than workers hired in the COA.

Migrant Health apparently became concerned about the worsening quality of ES-223 migrant and seasonal worker data and attempted in the mid-1980s to develop a production-based estimation procedure. Unfortunately, its efforts to support a production-based approach failed, so that Migrant Health apparently allocates funds to clinics on the basis of clinic administrative data. While convenient, such a service approach is not sensitive to changing migration patterns: migrants in many areas of California, for example, have settled in the area but may continue to use clinic services and thus generate a continuing "need" for funds, while migrants in remote areas that have recently begun to produce labor-intensive commodities may have no health services available because there is no clinic in the area to generate service data.

THE MID-1970S REPORTS OF LILLISAND AND RURAL AMERICA

Migrant Health produced two migrant estimates based on ES-223 data and one based on production data, but apparently began to de-emphasize these studies in the 1980s to allocate funds in favor of administrative clinic data. The mid-1970s report of Lillisand, by contrast, continues to guide the allocations of Legal Services Corporation (LSC) funds to Migrant Legal Action Programs, even though the Lillisand et al., study reached sharply different conclusions about the number and distribution of migrants than a similar mid-1970s Rural America study. If the Migrant Health experience illustrates how inconsistent studies get de-emphasized over time, the Lillisand study illustrates how even a study acknowledged to be flawed can continue to guide allocations because of grantee preferences for continuity in funding.

The Lillisand study was commissioned by LSC in 1976 to determine the number of migrants in need of legal services and their state-by-state distribution. The study consists largely of a critique of farm labor data sources and an extensive bibliography. Lillisand defined a migrant as a person who "left home temporarily overnight to do hired field or food processing work with the expectation of eventually returning home" (p. 50).[3] Lillisand's definition differs from the CPS-HFWF definition because it does not require a worker to cross a state or county boundary and it includes food processing workers. Lillisand also estimated the number and distribution of seasonal farmworkers; seasonal workers were defined as those who did less than 250 days of farm or food processing work and who did not leave home overnight.

The Lillisand estimates were based on four state-by-state data sources: a mail survey of over 600 public and private organizations serving migrants (about 50 percent responded); DOL ES-223 In-Season Farm Labor data for 1976; Migrant Health farmworker estimates for 1973; and Migrant Education data on children whose records were in the Migrant Student Records Transfer System (MSRTS). The report notes that the presumed reliability of the data follows this same ranking, i.e., Migrant Education is considered least reliable. However, in a few instances, such as Kansas, a lower Migrant Health estimate was used instead of ES-223 data because "Migrant Health Programs were more in touch with the population of concern to LSC" (p. 51). Most state estimates were based on ES-223 and Migrant Health data that was adjusted to reflect the survey results.

The Lillisand questionnaire asked farmworker service organizations to estimate the number of migrant and seasonal workers each month in the area served. Next, Lillisand requested an unduplicated annual estimate of migrants and seasonals, the definition of migrant and seasonal worker used to make these estimates, and the number of dependents. The questionnaire asked assistance organizations about the state-of-origin of migrants, the state from which migrants moved into the area, and the state to which they moved. Finally, respondents were asked about the race/ethnic background of farmworkers.

Surveys of farmworker service organizations which request data on the number of farmworkers often obtain some version of ES-223 migrant jobs data, as Lillisand noted on page 40. The Lillisand questionnaire requested additional farmworker characteristics data, but it did not ask survey respondents to distinguish single and family migrants and it did not list "white" or "Asian" as a race/ethnic choice for workers in the area, just Black, Spanish-speaking, Native American, and other.

Lillisand apparently estimated the number of migrants and dependents in each state on the basis of ES-223 data and, if the "adjusted ES-223 estimate" agreed with the survey responses, it was adopted. If there was a discrepancy between the adjusted ES-223 estimate and the survey responses, the 1973 Migrant Health estimates were used. When ES-223 (and sometimes Migrant Health) data indicated no migrants in a state, Migrant Education data on the children of both migrant and settled-out workers were used to generate state estimates. Settled-out or non-migrant children and their parents were included because "there is substantial doubt that the majority of these children maintain their settled out status" (p. 53).

The ES-223 reports define migrants as a subset of seasonal farmworkers. Seasonal farmworkers are defined in the ES-223 reports as persons who did farmwork for wages at least 25 days during the preceding 12 months but not more than 150 consecutive days with any single farm employer (unless the individual is "hired repeatedly on a short-term basis" by one employer).

Lillisand made no attempt to adjust migrant estimates for the duplication which occurs when a worker is counted on one farm one day and another the next day of the survey week because "while some workers may be counted more than once during a state's peak month, the migrants would have the same use for service wherever they travel" (p. 53).

Lillisand's adjustment of ES-223 data for each state should be a straightforward extrapolation from the published reports of peak migrant activity. No sample computations are included in the report, but it is not possible to reproduce the California estimate in accordance with the procedure outlined by Lillisand (p. 51). For example, the ES-223 September 1976 peak 54,350 migrant worker estimate in California should convert to 244,949 reported migrants and dependents according to the following procedure:

1. Multiply peak month migrant employment by 1.25 to account for worker turnover: September 1976 peak migrant employment = 54,350 x 1.25 = 67,938 migrant workers.
2. Assume two workers per household and divide this number of migrant workers by two to estimate 33,969 households; Lillisand assumed two migrant workers per household in 19 states with Mexican-American influxes and 1.4 migrant workers per household elsewhere.
3. Separate these households into 78.6 percent families and 21.6 percent one person households and multiply the family households by an assumed 6.65 dependents each, or 33,969 x 0.784 = 26,632 households x 6.65 = 177,099 migrant dependents.
4. Add 67,938 migrant workers and 177,099 migrant dependents to estimate 245,037 migrant farmworkers and dependents in California. This estimate was increased by the 292 certified H-2 workers in California and an assumed 15 migrant food processing workers for every 100 migrant field workers (0.15 x 67,938 = 10,191) for a total of 255,520 migrants and dependents.

The estimate computed here is 4 percent higher than Lillisand's estimate, probably because Lillisand did not specify which of household separation factors selected from the several available in the 1970 Census of Population was used.

Lillisand adjusted ES-223 data in this fashion for each state. If the resulting state estimate of migrant workers and dependents "differed significantly" (p. 52) from the survey data obtained from farmworker service organizations, the 1973 Migrant Health Target estimates were used instead. The adjusted ES-223 data was adopted for 23 states, and these states were assumed to have 560,000 migrants and dependents or 36 percent of the 1976 U.S. migrant population.

The 1973 Migrant Health Target estimates were adopted for 20 states and Puerto Rico. These Migrant Health estimates were based on local clinic estimates and ES-223 data and then adjusted in a fashion similar to Lillisand. Migrant Health estimated 700,000 migrant workers and dependents in 1973, or 900,000 fewer than Lillisand reported in 1976 even though Lillisand justified using Migrant Health data because "the actual numbers [of migrants] had not changed greatly since the Migrant Health Study was conducted" (p. 52).

One of the most puzzling features of the Lillisand study is the substantial difference between the published Migrant Health estimates and the numbers Lillisand reported for the 20 states in which "1973 Migrant Health figures were used" (p. 52). Quick perusal of the table reveals that Lillisand inflated the Migrant Health numbers, although there is no explanation for the 91 percent increase.

Thus, Lillisand adopted the Migrant Health estimates but increased them. Migrant Health reported 412,375 workers and dependents in the 20 states where Lillisand reported using this Migrant Health data; Lillisand reported 786,274 migrants and dependents in these states, or a 91 percent increase in three years. Lillisand does not explain whether or how the Migrant Health data was "adjusted," but the increase could result from if Lillisand changed the "dependents per migrant worker" factor--Migrant Health assumed 1.5 to 2.7 dependents per migrant worker, while Lillisand assumed 3.2 dependents per worker in states with a "Mexican-American influx." The 91 percent increase in the migrant population between 1973 and 1976 is very questionable; the HFWF, for example, reported a statistically insignificant 5 percent increase over this period.

Lillisand adjusted Migrant Education data in unspecified ways in four states (Mississippi, Maine, Nevada, and South Dakota) for which ES-223 reports indicated no migrants. Lillisand justified the inclusion of the parents of settled-out migrant children because he asserted that ex-migrants will become migrants again, even if Migrant Education data indicated that they did no migrant farmwork during 1976. Migrant population estimates for three states (New York, North Carolina, and Oregon) were based on farmworker service program estimates, but the basis of these estimates was not disclosed.

The Lillisand estimation procedure has several flaws. First, Lillisand does not report workers and dependents separately, so it is hard to link the raw job and worker estimates from which the reported population totals are derived with Lillisand's published numbers.

Second, the mixture of methods used to estimate the migrant population of each state makes it difficult to determine just what assumptions and factors were used to adjust base data such as the ES-223 data. Unlike Rural America, Lillisand only described the procedure for adjusting ES-223 data; there is no explanation of why Lillisand's Migrant Health estimates do not agree with those published in 1973. Similarly, there is no explanation of how Migrant Education data was adjusted or what methodology was used by the three farmworker service programs to estimate migrants.

Third, Lillisand is the only study reviewed which made no adjustment for duplication in the ES-223 data. Lillisand took peak month employment in each state and increased it by 1.25 to account for worker turnover in order to correct for two workers being hired to fill one job slot during the survey period, but Lillisand did not take the next step and correct for the duplication which occurs if a worker is employed on two farms during the survey period. Since duplication offsets turnover in most studies, Lillisand's worker estimates are generally 25 percent higher than those in other studies.

Fourth, Lillisand uses very high estimates of workers and dependents per migrant household: 2.0 workers and apparently 6.65 dependents. The 6.65 dependents assumption is not clearly specified. Lillisand says (p. 42): "a family size of 6.65 with two workers per family was applied to obtain the number of dependents per household" in the 19 states with a predominantly Mexican-American influx. It is not clear if Lillisand means that the average number of dependents per migrant household is 6.65 - 2.0 = 4.65 or if 6.65 is the dependent assumption. However, if 4.65 dependents per household factor is substituted into the calculation for California, the migrant population is only 202,260 instead of Lillisand's reported 244,949. Thus, it appears that Lillisand assumed 3.22 dependents per migrant worker in 19 states, or that over three-fourths of the migrant population in such states was comprised of nonworking dependents.

Although Lillisand was apparently comfortable with such a high dependents per migrant worker factor (on page 28, he notes that one can assume five dependents per worker), it appears to be too high. The 1980 Census of Population for California reported that only 6.5 percent of the "Spanish origin" persons lived in rural areas and that rural Hispanics averaged 4.0 persons per household and 4.3 persons per family. The 1983 HFWF sample had 1.1 migrant workers and 1.95 dependents per migrant worker household, and a 1983 survey of largely Hispanic farmworker households in California reported 1.75 workers and 1.85 dependents per household, or 3.6 persons per household, slightly less than the average reported in the 1980 COP for rural Hispanics. Substituting the 1.85 dependent factor into the California calculation reduces Lillisand's migrant population by 50 percent. If Lillisand's total estimate for all states (excluding Puerto Rico) of 1,511,341 is reduced by 50 percent to 755,670, it is much closer to the 1976 estimate of Rural America (649,432) and the 1973 estimate of Migrant Health (653,032) for the 50 states.

Rural America prepared a report in 1977 entitled "Where have all the farmworkers gone?" which concluded that farm labor data sources systematically undercount migrant and seasonal farmworkers. Rural America generated state-by-state estimates of migrants for 1976 in a fashion very similar to Lillisand, viz., Rural America's migrant estimates were based on 1976 ES-223 data and 1973 Migrant Health estimates. Rural America also "corrected" this data for perceived errors and omissions, and concluded that there were 236,000 migrant workers and 414,000 migrant dependents in 1976.

Rural America's migrant worker and dependent estimates are considerably lower than Lillisand's. Rural America took the peak-month migrant estimate from ES-223 data and increased it by 25 percent to account for worker turnover. However, migrants who change employers frequently could also be counted twice during the ES-223 survey period, and to minimize such duplicate counting, Rural America multiplied the peak month estimate by 1.25 and then reduced it by 0.25. For example, the ES-223 report estimated that a peak 54,350 migrants were employed in California in September 1976. Rural America first increased this peak month estimate by 1.25 (to 67,938) to account for turnover and then reduced it by 0.25 (to 50,954) because of possible duplicate counting.

Rural America then divided this adjusted migrant worker estimate into households based on the 1975 HFWF finding of 1.4 migrant workers per household, so 50,954 ÷ 1.4 = 36,396 migrant households in California. Rural America separated migrant households into single person and family households based on the ratio of single person to family households in rural areas in the 1970 Census of Population (all family households, not farmworker households). For California, 20 percent or 7,316 households were considered one-person migrant households and 29,080 family migrant households were assumed to have the same family size as the average for poor rural families in the 1970 COP. Thus, California's 29,080 family migrant households were multiplied by 3.63 dependents to obtain 105,542 dependents, and California's total migrant population estimate was 50,964 workers plus 105,542 dependents or 156,496.

Lillisand and Rural America used the same basic methodology to estimate migrants in 1976, viz., adjust ES-223 and Migrant Health data. However, very different migrant counts and distributions were generated: Lillisand estimated 1,511,341 migrants and dependents in 48 states; Rural America estimated 648,898 (both reported no migrants in Alaska and Hawaii). The major reasons for Rural American's lower migrant estimate is that Rural America adjusted ES-223 data to account for duplication and that Rural America used much lower dependents-per-migrant-household factors than Lillisand.

The three mid-1970s bottom-up migrant studies are compared in Table 3.3. Note the similarity between the migrant population estimated by Migrant Health Rural America (about 650,000), but their very different distributions of migrants across states. States with very different phases of migrants include Texas, California, and Florida; Migrant Health estimated that they included almost 50 percent of all migrants, and Rural America only 35 percent.

CENTAUR FARMWORKER STUDIES FOR OSHA

Centaur Associates prepared several reports for OSHA on the number of workers likely to be affected by a March 1, 1984 proposal to require the provision of toilets, drinking water, and hand washing facilities to hand laborers in the fields of farm employers with 11 or more such workers on one day (Federal Register,

Table 3.3
Bottom-Up Migrant Studies for 1973 and 1976

State	Migrant Target Population Estimates for 1973		State	Lillisand Mig Pop Estimates for 1976		
	Mig Pop	Per Dist		Mig Pop	Source	Per Dist
Texas	153,731	24%	Texas	318,225	DOL	21%
California	83,233	13%	California	244,949	DOL	16%
Florida	76,450	12%	Florida	166,964	MH	11%
Michigan	51,776	8%	Michigan	77,664	MH	5%
Washington	28,309	4%	Washington	70,743	DOL	5%
Minnesota	25,193	4%	Ohio	48,806	DOL	3%
Illinois	24,247	4%	Minnesota	43,457	MH	3%
Ohio	19,433	3%	Illinois	41,826	MH	3%
Oregon	16,749	3%	Oregon	41,431	FSP	3%
Idaho	14,462	2%	North Carolina	40,250	FSP	3%
Top Ten States	493,583	76%	Top Ten States	1,094,315		72%
United States	653,032	100%	United States	1,511,341		100%

Where Have all the Farmworkers Gone?
Rural America estimates for 1976

State	Mig Workers	Per Dist	Mig Dependents	Mig Pop	Per Dist
California	50,954	22%	105,542	156,496	24%
Washington	15,884	7%	21,201	37,085	6%
Florida	15,044	6%	30,254	45,298	7%
Ohio	14,215	6%	19,440	33,655	5%
North Carolina	13,841	6%	30,304	44,145	7%
Michigan	10,355	4%	13,280	23,635	4%
Indiana	9,194	4%	11,601	20,795	3%
Georgia	8,535	4%	18,901	27,436	4%
Texas	7,454	3%	19,433	26,887	4%
Minnesota	7,115	3%	10,260	17,375	3%
Top Ten States	152,591	65%	280,216	432,807	67%
United States	235,563	100%	413,869	649,432	100%

Sources: Migrant Health Program Target Population Estimates, mimeo, 1973
Lillisand, D. et. al. An Estimate of the Number of Migrant and Seasonal
  Farmworkers in the United States and Puerto Rico, 1977
Lillisand source is the source of the bottom-up data, e.g. DOL is ES-223 data,
  MH are the 1973 Mig Health estimates, and FSP means farmworker
  service programs
Rural America, Where Have all the Farmworkers Gone?, 1976

March 1, 1984, pp. 7589-7605). The Centaur methodology is straightforward. Centaur first "identified" the types of farms likely to employ 11 or more hand laborers (farms growing fruits and vegetables plus sugar, cotton, and tobacco). Centaur then distributed the production of these crops across states based on Agricultural Statistics acreage and yield data. Third, Centaur determined the minimum acreage necessary to employ 11 or more workers (e.g., 100 acres of vegetables, citrus, or grapes and 15 acres of tobacco). Fourth, the estimated number of farms and production was extrapolated from 1978 Census of Agriculture data, e.g., the 1,905 farms with 100 or more acres of citrus are only 8.3 percent of all citrus farms but produce an estimated 72.3 percent of all citrus (p. 20). Fifth, the distribution of 1981 hired workers across crops from the HFWF report was multiplied by the percentage of farms likely to employ 11 or more workers to yield 130,000 affected workers in tobacco, 215,000 in vegetables, and 182,000 in fruits, or a 527,000 total workers. Finally, these affected-hired-worker totals are multiplied by 0.392 to reflect the HFWF finding that the average hired worker is employed 92 days, or 0.392 of the Centaur-stipulated 250 day man-year. This 0.392 factor times 527,000 yields 207,000 affected man-years of farmwork. Centaur concluded that 527,000 of an estimated HFWF total 2.5 million hired workers in 1981 are employed on 52,000 farms that have 11 or more hand field workers on one or more days and would be affected by the OSHA proposal.

Centaur generated a variety of farmworker estimates by state and region. For example, Centaur distributed the 527,000 OSHA-affected hired workers as follows: California 34 percent; Florida 15; Southwest 5; Northwest 10; Upper Midwest 8; Southeast 24; and the Northeast 4 percent (p. 25). The percentage of migrants is derived from unspecified "case studies" California had 35 percent migrants; Florida 77; Southwest 75; Northwest 61; Upper Midwest 61; Southeast 38; and the Northeast 43 percent (p. 29). Centaur defined a migrant as a farmworker "who lives most of the year outside the county in which he/she is performing farmwork" (p. 29). Note that the number and distribution of affected farmworkers is derived from the Census of Agriculture, Agricultural Statistics, and the HFWF, but the percent migrant in each state and region is based on "Centaur case studies" in which migrants range from 0 to 95 percent of the farm workforce.

Centaur derived much of its demographic data from the CPS-HFWF report, and accepted the HFWF's estimate of 115,000 migrants in 1981. Centaur assumed that all of the HFWF's 115,000 migrants worked in its "hand-labor crops" in 1981, implying no migrants in livestock or grains, so that 115,000 migrants are 22 percent of the total 529,000 OSHA-affected migrants (p. 34). However, Centaur then says that each migrant is likely to work in three separate counties during the year, implying at least three jobs per migrant, so that in a "county-count" which assigns one weight for each worker in each county--regardless of duration--414,000 seasonals plus 115,000 x 3 = 345,000 totals 759,000 "county-counts". Somehow, Centaur concludes that such a county-weighting scheme makes migrants 52 percent of peak affected employment, although 345,000 ÷ 759,000 = 45 percent.

The Centaur study adjusts both regularly published HFWF data and case study data. Its production-approach is similar to that of HCR to determine the distribution of farmworker activity. However, instead of HCR's multiplying acres times hours, Centaur simply takes the actual production of the crops it identifies as labor-intensive. Centaur's national assumptions are troublesome: its estimated minimum acreage needed to employ 11 or more workers is simply not sensitive to real-life ownership and labor-arrangements. Since 80 to 90 percent of California-Arizona citrus, for example, has corporate or absentee owners, virtually all citrus is picked by hired harvesting crews which average 30 to 35 pickers. Furthermore, the assignment of a farm to a particular commodity category in the Census of Agriculture depends on its primary crop, so that a corporate livestock operation which also grows vegetables may be listed with livestock farms and not vegetable farms even if all of its vegetables are harvested by crews of hired workers.

Centaur's major assumption is that the HFWF count of farmworkers is reasonably accurate, that is, based on December 1981 CPS interviews with about 1,500 farmworker households, the HFWF estimated there were 2.5 million hired workers. These households reported their primary farmworker activity, and 34 percent worked primarily in grains and other field crops; 19 percent in livestock and dairy; 23 percent in fruits and vegetables; 11 percent in tobacco; and the remainder in cotton, nurseries, and other crops. Centaur's assumption is that the HFWF number and distribution of hired workers across crops accurately reflects reality. Centaur uses HFWF for the total number of farmworkers; their distribution across crops (but not across regions); and average days worked.

SUMMARY

This chapter reviewed three types of migrant farmworker studies. Most familiar are Migrant Health-type bottom-up studies, which begin with snapshot estimates of migrants at work during a particular week and then adjust ES-223 data to account for turnover, duplication, and survey deficiencies. Such bottom-up studies based on migrant snapshot surveys dominated 1970s migrant studies, but have become untenable in the 1980s because the quality of the already-poor ES-223 data has deteriorated. Nonetheless, the number and distribution of migrants in these 1970s studies still dominate many migrant discussions and are the basis for allocating some migrant assistance funds.

Some migrant assistance and regulatory agencies began to abandon this adjust-the-snapshot-migrant-estimate approach in the 1980s. Migrant Health tried to develop a procedure to estimate the number of migrant and seasonal workers on the basis of farm production data, but apparently abandoned this approach in favor of internal administrative data. OSHA also commissioned studies of the number and distribution of workers likely to be affected by regulatory rules which relied on the CPS household data collected in December, but this approach has been rejected by most other agencies.

NOTES

1. To put a 14 person-per-year-round job-turnover-ratio in perspective, if the U.S. Armed Forces had 14:1 turnover, about 28 million persons or the population of California would have to serve sometime during the year to maintain current troop levels.

2. The calculation is 100 migrants divided by 1.5 workers per household to yield 67 households. If 80 percent are families, 0.8 x 67 = 53 families, times five dependents each is 267 dependents. Adding 267 dependents and 100 workers yields a migrant population of 367 for the survey week.

3. Lillisand complained that migrants working in their home or base areas will be classified as seasonal workers (p. 45), implicitly refusing to recognize that a "migrant farmworker" may have several occupational titles during a particular time period.

Chapter 4

A Top-Down Procedure

There are several sources of bottom-up data on migrant
farmworkers which generate a count and distribution of migrants,
such as the 159,000 persons who did at least one day of migratory
farmwork in 1985 (CPS-HFWF); the 765,834 migrant man-months
of employment estimated by the ES-223 data in 1987; or the
600,000 children of migrant farmworkers in the MSRTS in 1982.
Each of these bottom-up data sources has serious deficiencies, so
this chapter presents a top-down method to estimate the number and
distribution of migrant farmworkers across states. It should be
emphasized that top-down estimation methods cannot make the
precise distinctions between individuals that are possible in
household surveys. The top-down method developed in this chapter
makes an important assumption: the state-by-state distributions of
migrant and seasonal workers are identical, meaning that exactly the
same method would be used to distribute both migrant and seasonal
workers across states, but not to estimate the number of seasonal
workers.

The top-down method of counting and distributing migrants
developed in this chapter relies on ES-202 or UI data to estimate the
number of migrants, and then it combines Census of Agriculture,
QALS, and CPS-HFWF data to distribute these migrants across
states. A major virtue of this approach is that the major data bases
on which it is based, namely UI and QALS data, are improving.

THE NUMBER OF MIGRANT WORKERS

The quarterly employment and wage data that farm employers report to Unemployment Insurance (UI) authorities is the source of data on the number of migrant workers. This UI data is assembled by the Standard Industry Code (SIC) or commodity of the reporting employers, and the county and state where the farm is headquartered is reported.

The UI data permit farm employers to be defined as employers whose SIC codes are those for crops (01), livestock (02), and selected agricultural services (07*). A person or Social Security Number (SSN) reported at least once by such a farm employer is a "farmworker," and migrants are a subset of all such farmworkers. Note that such farmworkers include supervisors, field and livestock workers, and persons with nonfarm occupations such as clerks and accountants. The migrant definition closest to the USDA-CPS-HFWF cross-county-lines-and-stay-away-from-home-overnight definition is one which requires farmworkers to have at least one farm employer outside their home or base county. Such a definition does not necessarily require an overnight stay away from home, as does the CPS-HFWF definition, but it comes closest to this definition and is thus abbreviated as the "HFWF" migrant definition; the quote marks signify that it is close to but not identical to the HFWF definition developed by USDA to analyze CPS data.

The UI data readily available on a state-by-state basis are the number of employers, average annual employment, and total wages paid by commodity and county. The advantage of UI data is that they are a "census" of covered employment and wages, not simply a sample. However, as administrative data, UI data are reported in order to collect the taxes needed to operate the UI system, not to enumerate migrant workers. This means that only the employees of "large" farm employers are included in most states, and even for them, what is reported is the number of year-round-equivalent workers, not the number of workers they actually employed, i.e., a reported average annual employment of 12 could reflect one worker hired year-round or 12 workers employed during one month.

California is a fortuitous exception to the limited UI coverage of farmworkers and the limited UI data available. In California, virtually all farm employers (those paying $100 in wages or more in a calendar quarter) must report the names, wages, and weeks worked by all persons on their payrolls. This means that a worker (SSN) reported at least once by an employer whose SIC code is 01,

02, or 07* is a "farmworker," and that all of the farm and nonfarm jobs of such farmworkers can be assembled.

Once the complete employment record of each person with at least one farm job is assembled, migrant farmworkers can be defined and counted. In 1984, there were 921,500 workers (SSNs) who were reported to California UI authorities at least once by a farm employer; and a one-percent random sample of 9,215 workers was drawn. Then various definitions of migrant were applied to these 9,215 workers in order to determine how many were migrants and the characteristics of these migrants (Table 4.1).

In the UI data, a migrant had to have at least two reported jobs in California. The "HFWF" definition was applied by examining the earnings from each job and assigning the worker to the home or base county in which he or she had maximum earnings from all farm and nonfarm jobs. "HFWF" migrants were persons who had at least one farm (01, 02, or 07*) job outside their base or highest-earning county. This "HFWF" migrant definition reinforces the well-known "salvage" theory of the harvest labor market, which holds that urban workers translate otherwise idle time into earnings by heading out to farms to pick crops.

About 20 percent of sample California farmworkers were "HFWF" migrants, meaning that they had at least one farm job outside their home or base county. Most of these "HFWF" migrants (84 percent) had just one farm job outside their base counties; 12 percent had two farm jobs outside their base counties; and 4 percent had three or more migrant farm jobs. The home or base counties of "HFWF" migrants are surprising: the top five home counties were home to almost half of the "HFWF" migrants, and two of these counties--Los Angeles and Orange counties--were home to about one-fourth of all "HFWF" migrants. Thus, the "HFWF" migrant definition suggests that about 20 percent of the workers employed on California farms hold at least one farm job outside their home or country of highest earnings from all jobs, suggesting that California has about 188,000 "HFWF" migrants.

Migrancy can also be defined in other ways. A migrant could be required to have at least two farm jobs in at least two counties. About 12 percent of the sample California workers were such "intra-ag" migrants, implying about 113,000 intra-ag migrants. Half of the intra-ag migrants were based (had highest earnings) in

Table 4.1

Migrant Farmworker Definitions from the 1984 California Unemployment Insurance Data

Total	Workers	Crops-01	Livestock-02	(2)Ag Services-07	Other 07	(3)Multiestab-07	(3)Multiestab Nonag	(4) Jobs
Sample	12,436	8,337	556	7,146		3,756	4,171	23,966
Per Dist-Jobs		35%	2%	30%		16%	17%	100%
Workers	9,215	8,337	556	6,154	992	3,728	4,171	23,938

Workers employed in 1 County	2	3	4	5	6	7	8	Migrant Workers

INTRA-AG MIGRANTS--1130 WORKERS OR 12.26 PERCENT WITH ONE 01,02,OR 07*JOB IN ONE COUNTY AND ANOTHER FARM JOB IN ANOTHER COUNTY--1918 MIGRANT FARM JOBS OR 1.7 JOBS EACH

9,215	8,085	888	184	40	14	2	1		1,130
--PER DIST	88%	10%	2%						12%
JOBS		7,974	513	5,919		0	0	14,406	
--PER DIST		55%	4%	41%				100%	

"HFWF" MIGRANTS--1873 WORKERS OR 20.33 PERCENT HAD AT LEAST ONE FARM JOB(O1, 02, OR 07*) OUTSIDE THEIR BASE COUNTY--2092 MIGRANT FARM JOBS OR 1.1 FARM JOBS EACH OUTSIDE THE COUNTY OF HIGHEST EARNINGS FROM ALL FARM AND NONFARM JOBS

9,215	7,342	1,579	224	49	17	2	1		1,873
--PER DIST	80%	17%	2%	1%					20%
JOBS		7,974	513	5,919		907	3,925	19,238	
--PER DIST		41%	3%	31%		5%	20%	100%	

TOTAL MIGRANTS--2201 WORKERS OR 23.88 PERCENT WITH AT LEAST ONE FARM JOB(01, 02, OR 07*) IN ONE COUNTY AND A FARM OR NONFARM JOB IN ANOTHER COUNTY--3676 JOBS OR 1.7 JOBS EACH

9,215	7,014	1,705	383	76	29	4	1	3	2,201
--PER DIST	76%	19%	4%	1%					24%
JOBS		7,974	513	5,919		907	3,925	19,238	
--PER DIST		41%	3%	31%		5%	20%	100%	

Source: Special tabulation of data from California crop, livestock and agricultural service employers; 1 percent sample of 1,243,951 unique SSN. THERE WERE 921,150 WORKERS WITH AT LEAST ONE IDENTIFIABLE FARM JOB, AND 187,300 WERE "HFWF" MIGRANTS IF THESE SAME MIGRATION PERCENTAGES ARE ALSO APPLIED TO THE MULTIESTABLISHMENT WORKERS, THEN THERE ARE 252,895 "HFWF" MIGRANTS IN CALIFORNIA--20.33%x1,243,951=252,895
(1)THE NUMBER OF FARMWORKERS WITH AT LEAST ONE 01, 02, OR 07*JOBS IN 1984(071,072,0741,0751, OR 076)
(2)THE NUMBER OF 07* JOBS HELD BY THE FARMWORKERS DEFINED IN (1)
(3) MULTIESTAB--2074 OR 16.7 PERCENT OF ALL SAMPLE FARMWORKERS WERE EMPLOYED ONLY BY MULTIESTAB ERS
(4) 28 WORKERS FILED CLAIMS FOR UI IN CALIFORNIA BASED AT LEAST IN PART ON OUT-OF-STATE WAGES

just five counties, and Fresno, Tulare, and Kern counties included almost 40 percent of all intra-ag migrants.

The third migrant definition is the most expansive. A"total" migrant was defined as any worker with at least two jobs in two counties, provided that one of these jobs was a farm job. About 24 percent of the sample satisfied this total migrant definition, for 220,000 total migrants. Total migrants are similar to "HFWF" migrants in being based in both urban and rural counties.

Suppose a farmworker has four jobs that pay the following wages: Fresno grapes ($800); Los Angeles hotel ($900); Los Angeles nursery ($600); and Tulare peaches ($400); or three farm jobs in three counties. The farmworker's base county for the "HFWF" definition is Los Angeles because it generates the worker's maximum earnings from all jobs. The worker is an intra-ag migrant because he/she has at least two farm jobs in two counties; a "HFWF" migrant because he/she has one farm job outside the base (Los Angeles) county; and a total migrant because the worker has one farm job in one county and another farm or nonfarm job in another county.

If a worker had two nonfarm jobs in Los Angeles--hotel ($600) and janitor ($1,000)--and one Fresno grape job, the worker is a "HFWF" and total migrant but not an intra-ag migrant. If a worker had a farm and nonfarm job in Los Angeles [nursery ($600) and hotel ($600)], and a Fresno hotel job ($500), then the worker is a total migrant but not an intra-ag or "HFWF" migrant because the job outside the base county is not a farm job. If the Fresno hotel job paid $1,500, then the worker would be a "HFWF" migrant because Fresno becomes his/her base county and he/she has a farm job in Los Angeles.

Under all three definitions, most migrants worked in only two counties. For example, 79 percent of the intra-ag migrants had farm jobs in two counties and 21 percent had farm jobs in three or more counties. Similarly, 78 percent of the total migrants had a farm job in one county and another job (farm or nonfarm) in a second county; 17 percent had other jobs in three counties; and only 4 percent had other jobs in four or more counties.

THE DISTRIBUTION OF MIGRANTS

Defining migrancy in terms of the UI "census" of persons employed on California farms yields 113,000 to 220,000 migrants, more than the peak 50,000 estimated by ES-223 data in the mid-1970s studies. To convert these California migrants into a national count and distribution of migrants across states, a top-down formula must be developed which indicates each state's share of migrant activity. Since the UI data has incomplete coverage outside of California, the top-down migrant formula must be based on other data sources.

One problem with past migrant studies is that they "adjusted" source data in an unreported fashion, making it impossible to replicate the formula in order to see if the distribution changed or how sensitive the formula is to particular adjustments. Thus, distribution formulae which adjust source data have an obligation to be as straight-forward as possible in explaining exactly what adjustments were made.

The distribution formula presented in Table 4.2 satisfies this easy-to-understand criterion. The formula begins with the labor expenditures of crop farms: how much did crop farmers report spending for labor they hired directly and through contractors in the 1982 COA? This assumes that most migrants are employed on crop farms, and that labor expenditures are a reasonable proxy for each state's volume of migrant farmworker activity.

The expenditures-as-a-proxy-for-activity assumption may be questioned because higher-wage states will generate greater labor expenditures. In order to account for state-by-state differences in hourly wages, each state's labor expenditures were divided by the average state- or region-wide hourly wage paid to fieldworkers in July 1982. Dividing labor expenditures by an hourly wage generates a distribution of hours-of-work done by hired workers on crop farms across states.

Hours of hired crop work are not a complete top-down indicator of migrant activity. A second top-down migrant indicator is how many workers crop farmers reported hiring to work less than 150 days or seasonally on their farms in the 1982 COA. As noted previously, this less-than-150-day measure indicates the volume of hiring, not necessarily migrancy, but there are generally more migrants in state's with a higher volume of seasonal hiring.

The final migrant indicator is one that is more of a bottom-up data source. The two data sources which report counts and distributions of migrants are the CPS-HFWF and ES-223 reports. Both have deficiencies: the CPS-HFWF is based on a very small sample, and the ES-223 represents incomplete and ad hoc guesses

Table 4.2

The Distribution of Farmworker Activity: 1982

State	COA Crop Labor Expend 1982-000	Hourly Field Wage "July-82"	Expendits Div by Wage Hours or Actv	Dist of Expendits Per Dist	Less than 150 day Crop Workers	Per Dist	CPS-HFWF Migrant Workers	Per Dist	Final Dist
California	1,681,436	$4.69	358,515	26%	607,601	22%	39,529	17%	23%
Florida	472,408	$3.90	121,130	9%	112,293	4%	42,664	19%	10%
Washington	262,910	$4.34	60,578	4%	205,624	8%	8,750	4%	5%
Texas	274,717	$3.93	69,903	5%	84,227	3%	15,177	7%	5%
North Carolin	183,623	$3.40	54,007	4%	207,800	8%	5,387	2%	4%
Georgia	90,000	$2.99	30,100	2%	49,795	2%	17,887	8%	4%
Kentucky	97,667	$3.09	31,607	2%	189,009	7%	3,496	2%	3%
Illinois	134,405	$3.53	38,075	3%	85,081	3%	4,890	2%	3%
Oregon	108,212	$3.66	29,566	2%	101,025	4%	4,657	2%	3%
Iowa	91,027	$3.45	26,385	2%	77,510	3%	4,241	2%	2%
Top Ten State	3,396,405		819,867	60%	1,719,965	63%	146,678	65%	62%
United States	5,409,270	$3.80	1,367,923	100%	2,730,046	100%	226,416	100%	100%

COA crop labor expenditures are the labor expenditures reported by crop farms in the 1982 COA

Hourly field wage is the average wage for the state in Farm Labor for July 1982; some of this data was revised in May 1985

Less than 150 day crop workers is the number of "seasonal" workers crop farmers reported hiring in the1982 COA

CPS-HFWF is the number of migrants in each region in the 1983 HFWF report distributed to states on the

basis of each states's crop sales in 1983

about migrants. The CPS-HFWF, despite its weaknesses, appears to be stronger than the ES-223 data, and is third indicator of migrant worker activity.

The CPS-HFWF migrant data is reported only for multi-state regions, and is not available for even-numbered years such as 1982. Thus, the 1983 CPS-HFWF regional estimates of migrants were assigned to the states within each region on the basis of each state's share of regional crop sales. For example, CPS-HFWF estimated that 19,000 migrants were living in the Southern Plains region which includes Texas and Oklahoma in 1983; Texas accounted for 80 percent of these two states crop sales, so 80 percent of these 19,000 migrants get assigned to Texas.

The standardized labor expenditures of crop farms, seasonal workers employed on crop farms, and migrant workers assigned to states are the three elements of the distribution formula. Given these formula elements, how should they be combined? There are no magic formula weights, but most farm labor data experts believe that COA labor expenditure data are the most reliable, so the standardized labor expenditures of crop farms are assigned a weight of one-half in the formula. Both employment indicators have deficiencies: the seasonal worker indicator includes all persons employed for wages, including family members, and it does not distinguish between a worker hired for one day on the responding farm and a worker hired for 149 days. The migrant indicator, on the other hand, enumerates migrants directly but does so with large sampling and nonsampling errors. Given these considerations, each employment indicator is assigned a weight of one-fourth.

The migrant distribution formula can be summarized as follows:

$$S_i = 0.5 \ [(\Sigma w_i) + H_i] + 0.25(E_i) + 0.25(M_i)$$

where $S_i =$ state's share of migrant activity.
$W_i =$ the labor expenditures and contract labor expenses of crop farms reported in the 1982 COA.
$H_i =$ fieldworker hourly earnings for state or region in the July 1982 Farm Labor.
$E_i =$ workers employed less than 150 days on crop farms in the 1982 COA.
$M_i =$ regional HFWF migrants in 1983 assigned to the states within each region on the basis of each state's share of the region's 1983 crop sales.

This labor expenditure, seasonal employment, and migrant formula distributes migrants across states in a pattern closer to Rural America than the other mid-1970s migrant studies. This distribution formula concentrates migrant activity in a handful of states, just as earlier studies did. Compared to e.g., the Lillisand formula, it redistributes migrant activity from Texas to California and North Carolina.

This migrant distribution formula can be combined with the migrant counts from California UI data to generate an estimate of migrant workers in each state. California has 23 percent of U.S. migrant activity according to this distribution formula, and thus there are 490,560 intra-ag migrants in the United States; 814,155 "HFWF" migrants; and 955,510 total migrants if the California estimates are expanded to the United States. Because of a quirk in the UI data, the county and commodity of employers with several sub-units are not recorded, so if the same migrancy percentages apply to the workers reported by such multi-establishment employers, U.S. migrant counts are higher, viz., 601,155 intra-ag migrants; 996,526 "HFWF" migrants, and 1,170,929 total migrants.

The obvious question raised by this method of counting and distributing migrants is how this procedure can generate a number of migrant workers that is so much higher and a distribution so much different from the mid-1970s studies. The higher migrant count-- 600,000 to 1.2 million versus 125,000 to 250,000 in the mid-1970s--probably reflects the undercounting of workers employed on farms in the ES-223 and other data which are typically used in bottom-up migrant studies. This undercount appears to be particularly severe for workers employed by agricultural service firms, and since the wages paid by these firms are almost equivalent to the wages paid by crop farms, missing their employees has serious ramifications. The different distribution probably better reflects migrant activity where migrants work; the mid-1970s distributions were based on ad hoc assumptions, so there is no reason to believe that they ever reflected reality.

ALTERNATIVE APPROACHES

This top-down approach to counting and distributing migrant farmworkers has been criticized, especially by migrant assistance groups whose funds would be re-shuffled by switching from a bottom-up to a top-down formula. The critics emphasize that top-down data does not isolate migrants; instead, they note that distributing migrant worker activity across states on the basis of labor expenditures assumes that migrants are distributed pretty-much as non-migrant farmworkers. At the state-level, this is a defensible assumption: most crop agriculture which hires seasonal workers also hires migrant workers, so the expenditures of crop farms on seasonal workers should be highly correlated with expenditures on migrant workers.

The major justification for using a top-down approach to count and distribute migrants is that there is no credible alternative. Bottom-up data is seriously flawed, making it virtually impossible to adjust or correct. Indeed, correcting or adjusting bottom-up data implies that there is a better data source available which can be used to correct it; logic dictates that this better data be used from the beginning.

Data from migrant assistance programs is sometimes offered to correct bottom-up base data. However, such client data cannot be used to adjust base data because clients simply reflect the distribution of funds, especially if assistance programs have about the same funding-client ratios in each state. Since all migrant assistance programs serve only a fraction of eligible migrants, correcting base data with client data or using client data directly to allocate migrant activity simply continues past funding patterns. Furthermore, the client data (with the possible exception of the MSRTS) simply reflects past (flawed) distributions, so using client data to indicate the distribution of migrants simply perpetuates the flaws of previous distribution studies.

SUMMARY

A two-part formula can generate the number and distribution of migrants. Migrants can be defined in several ways and their number estimated by "tracking" workers through the Unemployment Insurance System. Such a procedure yields 113,000, 188,000, or 220,000 migrants in California, depending on the exact definition; the middle figure, which defines a migrant as a person with at least one farm job outside the county in which he or she had their maximum earnings from all farm and nonfarm jobs, to be most reasonable.

The distribution of migrants across states is based on state-by-state labor expenditure data, hourly earnings, the reported employment of workers hired less than 150 days, and an estimate of migrants. According to this distribution formula, California has 23 percent of U.S. migrant activity, Florida 10 percent, and Washington and Texas each 5 percent, or the top 4 states included about 43 percent of all migrant activity in 1982. California's 23 percent share implies that there are about 1 million migrant workers in the U.S.

This combination of counting migrants with UI data and distributing them with labor expenditure and other data has been criticized using too many proxies to get at the migrants of interest. However, there is no credible alternative source data nor is there a clearly preferred formula to combine the most reliable data in a different manner.

Chapter 5

Migrant Farmworker Characteristics

Migrant farmworkers are persons who leave their home county and stay away from their normal residence overnight to do farmwork for wages. This inclusive county-crossing and overnight stay definition means that the heterogeneous group of persons who share the migrant label may have very little in common: some do one, some 100, and some 300 days of migratory farmwork annually. Some migrants change occupational titles several times during the year as, for example, they switch from nonfarm worker to local farmworker to migrant. Some migrants travel with their families and some travel alone; some family migrants have spouses and dependents who are also migratory farmworkers, but in other migrant families accompanying family members are nonfarm workers or not in the labor force.

Migrant farmworkers are a much misunderstood part of the farm labor force. Varden Fuller, a long-time student of farm labor, observed recently that many people "refer to persons who harvest crops as "migrants" without knowing. . . whether they were migrants or not" (Emerson 1984, p. xi). Fuller, who was Staff Director of the 1950 Presidential Commission on Migratory Labor, went on to note that the most significant farm labor change in the past three decades "is the decline in migratoriness" . . . [both] the decline in physical magnitude [and] the decline in the myth" (p. xi). Nonetheless, as Fuller notes, the migratory streams whose heavy arrows picture migrants moving south to north with the harvest from Florida, Texas, and California still capture public attention.

Books such as <u>Sweatshops in the Sun</u> (1973) and <u>A Caste of Despair</u> (1981) and documentaries such as <u>A Harvest of Shame</u> (1960) describe migrants as minority families who each year "follow the sun" south-to-north harvesting fruits and vegetables. Migrants are usually described as hard-workers who earn low wages and are frequently victimized by labor contractors or farm employers, making them one of the most disadvantaged groups in the United States. These exposé reports often report that there are "millions" of migrant farmworkers in the United States. For example, Goldfarb quotes a 1980 National Association of Farmworker Organizations speaker who asserted that "each year as the harvest begins, thousands of buses and cars haul thousands of crews to fields across America as millions of migrant farmworkers hit the road" (p. 9).

The <u>Wall Street Journal</u> reported in 1986 that there were two million migrants in the U.S., including 150,000 mostly (80 percent) Mexican-Americans in Florida (June 26, 1986, p. 1). Reporter Sonia Nazario went on to note that migrant life today is similar to that of 20 years ago: low earnings (average $7,000); few recent pay increases; and problems with peonage, child labor, and substandard housing.

The Lillisand report also includes many of these familiar aphorisms about migrants. Lillisand described migrants as having the characteristics of persons most likely to be missed in the decennial census of population, viz., ". . . poor, adult, male, young, and nonwhite" (p. 9). Migrants' ". . . traditional quarters as well as their unconventional but not unusual quarters (such as abandoned buses and autos, condemned buildings, abandoned farmhouses and chicken coops) were excluded from lists of residences" (p. 10) used to locate people by the COP. Furthermore, Lillisand asserts that April is not an appropriate month to survey migrants, because "while on the road in April migrants are most apt to be sleeping in unconventional quarters such as their autos, buses or trucks, or under the starry sky" (p. 12). Finally, filling out a census questionnaire "requires the ability to read, and to read critically."

Lillisand makes a variety of other observations on migrants while discussing data sources. Lillisand questions the HFWF finding that over half of all migrants are white (pp. 18-19); suggests that there may be an average five dependents per migrant worker (p. 28); and reports that "there are indications that migrants do not stay settled out for very long" (p. 35). Lillisand notes that an estimate of peak month migrant employment "undercounts by one-fourth to one-third the true number of migrants because of labor turnover" (p. 37). A summary conclusion in the Lillisand report captures its opinion of migrant data: "the best data [is] defined as that which more closely resembles reality" (p. 39). However, Lillisand's "reality" appears to be what is reported by farmworker assistance personnel, not the various statistical data sources.

The Lillisand summary contrasts sharply with USDA-HFWF migrant data, which reports that the number of migrants stabilized at 200,000 in the 1970s; that only one-third of the migrants are primarily farmworkers; and that many migrants are young White males who travel less than 200 miles to be migrant farmworkers. According to HFWF data, only one-third of all migrants worked on fruit, vegetable, and horticultural farms in 1983, while 55 percent worked on field crop farms and 10 percent worked on livestock farms. It is very difficult to reconcile the different pictures of migrants painted by exposés and HFWF reports.

This chapter reviews the Lillisand data on migrant farmworker characteristics, the HFWF migrant data, and data from a variety of other sources. In each case, the focus is on migrant family characteristics, employment and earnings patterns, and perceptions of the number of "career" migrants.

LILLISAND MIGRANT CHARACTERISTICS

The Lillisand report includes a 22-page discussion of migrant racial-ethnic characteristics, wages, age and sex, and migration patterns which is typical of most migrant studies Lillisand asserted that migrants ". . . move from place to place, live in labor camps and other temporary quarters, and often lack both knowledge of the existence and location of legal services programs and their own transportation to seek help" (p. 3). Thus, "migrants have special access problems [to legal services] distinct from those of seasonals and the nonfarm working poor" (p. 3). Lillisand reviewed migrant characteristics to develop procedures that would assist in the effective delivery of legal services to migrants (p. 76).

The Lillisand report notes that the 1975 HFWF data indicated that migrants were 63 percent White, 24 (HFWF p. 5 says 25) percent Hispanic, and 12 percent Black (p. 76). The HFWF data is rejected in favor of a 1975 Centaur Study done for OSHA, which estimated migrants to be 20 percent White, 50 percent "Spanish speaking," and 30 percent Black (p. 77). Lillisand then asserted that the 50 percent Spanish-speaking migrant worker estimate implies 750,000 Spanish-speaking clients, thus assuming (1) that there are 1.5 million migrants and dependents and (2) that all persons in a household with an Hispanic migrant worker speak Spanish.

Lillisand's brief discussion of wages presents 1970 income data and 1975 earnings data. Low farmworker earnings are attributed to the seasonality of farm employment and agriculture's exemptions from minimum wage and other labor laws (p. 78). Lillisand makes a series of assertions (p. 80) that suggest legal services attorneys must make a considerable outreach effort because migrant farmworkers live "a hand to mouth existence," are paid in cash, earn piecerate wages only when they work, and do not usually have adequate insurance and other benefits (p. 80). Lillisand apparently believed that all migrants were eligible for legal services aid, since there is no hint that some migrants' incomes may exceed the cut-off for subsidized legal assistance.

Lillisand asserts that migrants are primarily young and male, "partially because the average life expectancy of migrant workers is 49 years" (p. 80). Migrants "are forced into the labor stream," and Lillisand asserts that some migrants from Texas travel "as much as 12,000 miles in search of short-term jobs which pay the minimum wage, if that" (p. 81). [This is equivalent to four round trips between south Texas and Ohio.] Lillisand notes that some migrant attorneys are frustrated by their clients unwillingness to pursue legal remedies because of the migrants' desire to get out of the migrant stream, but "the fact of the matter is that the vast majority of them will return year after year for lack of alternatives" (p. 82), suggesting that migrancy is an unwanted career.

Lillisand cites Fuller's comment that maps which show streams of people moving south to north embody more flows of imagination than people but writes that such migration patterns continue to be reported because they are "based on sheer numbers of workers appearing in adjacent areas according to a seemingly logical sequence" (p. 82). Lillisand reviews data which suggest that traditional follow-the-sun migration has declined, and concludes that there has been "a continuing deterioration of traditional streams" (p. 88).

In sum, Lillisand concludes that half of all migrants are Spanish-speaking (all Hispanics are assumed to be Spanish-speaking), that migrant earnings are low, that migrants travel great distances, and that traditional south-to-north migration patterns were eroding in

1975. Many of these assertions are presumably based on Lillisand's survey of migrant assistance organizations.

CPS-HFWF MIGRANT CHARACTERISTICS

The CPS-HFWF data provide a sharp contrast to Lillisand's summary of migrant characteristics. The HFWF reports that the number of migrants stabilized at about 200,000 after 1970; that less than 40,000 migrants were Hispanics in 1983, when all migrants had average earnings of almost $6,000; and that less than one-third of all migrants in 1983 travelled 1,000 miles or more to do migrant farmwork (Table 5.1). The HFWF data is drawn from a very small but scientifically constructed sample; the Lillisand conclusions are based on a variety of reports and observations.

The HFWF data indicate that the number of migrants declined sharply in the 1960s, from about 400,000 in the early 1960s to 200,000 in the 1970s. This decline in migrant numbers matched the overall decline in hired farmworkers, so that the migrant proportion of the workforce remained stable at about 10 percent.

HFWF reporting makes it difficult to chart trends in migrant characteristics. In 1973, for example, the HFWF reported that Whites were 96 percent and "Negro and other races" were 4 percent of all migrant farmworkers. Migrants who were White worked an average 150 days and earned $2,500, and two-thirds of these days and earnings came from farmwork. The distances travelled by migrants were not recorded in 1973, but one-fourth of the migrants in that year were very young (between 14 and 17) and over half were younger than 25.

By 1983, the minority component of HFWF migrants had increased from 4 to 55 percent, although Whites (45 percent) and Blacks (39 percent) overshadowed Hispanics (15 percent); the White component of the total farm workforce, by contrast, remained stable at about 75 percent. Males comprised a stable 80 percent of the migrant work force, but the percentage of 14 to 24 year-old migrants fell from over half to one-third by 1983.

Migrant earnings were reported in 1983 for two groups: the estimated 124,000 migrant farmworkers who were household heads and the 103,000 migrants who were other household members. Migrant household heads had a median $6,045 in total earnings, and 81 percent of these migrant earnings were derived from an average 154 days of farmwork. Migrants who were not household heads

Table 5.1
Migrant Farmworker Characteristics in the December CPS-HFWF: 1960-1983

Characteristic	1983	1981	1979	1977	1975	1973	1971	1969	1967	1965	1963	1960
All Migrants	226,000	115,000	217,000	191,000	188,000	203,000	172,000	257,000	276,000	466,000	386,000	409,000
-White	45%	77%	63%	58%	64%	96%	91%	92%	83%	78%	66%	48%
-Hispanic	15%	17%	27%	31%	24%	-	-	-	-	-	-	25%
-Black	39%	8%	11%	11%	12%	4%	9%	8%	17%	22%	32%	-
Age 14-24	35%	25% *	55%	55%	64%	53%	64%	64%	52%	52%	45%	48%
-25-54	59%	15%	40%	37%	35%	41%	28%	31%	34%	37%	50%	42%
-55 and over	6%	1%	5%	9%	2%	6%	9%	6%	14%	11%	5%	10%
Male	86%	83%	76%	80%	76%	84%	83%	78%	74%	72%	82%	77%
Education-Pers >25	148,000	115,000 **	97,000	-	-	-	-	-	-	-	-	213,000
-Less than 8 years	66%	34%	38%	-	-	-	-	-	-	-	-	66%
-12 or more years	29%	40%	55%	-	-	-	-	-	-	-	-	12%
Annual tot earnings-$	5,921	3,995	4,852	3,761	3,324	2,587	1,630	1,732	1,424	1,362	821	875
-Farm earnings-$	4,623	2,729	2,277	2,263	2,000	1,688	1,628	1,729	922	795	486	650
--Per of tot earnings	78%	68%	47%	60%	60%	65%	100%	100%	65%	58%	59%	74%
-Days of farmwork	124	112	87	97	95	102	118	129	85	82	82	98
Daily farm earnings-$	37	24	26	23	21	17	14	13	11	10	6	7
--Hourly(div by 8)-$	4.66	3.05	3.26	2.90	2.63	2.07	1.73	1.68	1.36	1.21	0.74	0.83
--Fed min wage-$	3.35	3.35	2.90	2.30	2.10	1.60						
Migs who only did far	48%	72%	45%	52%	60%	53%	53%	57%	62%	54%	66%	47%
Poverty line-farm of 4	10,178	9,287	7,412	-	5,500	-	-	-	-	-	-	-
Census reg-Northeast	-	4%	55%	-	-	-	-	-	-	-	-	1%
-Midwest(Northcentral	-	21%	26%	-	-	-	-	-	-	-	-	8%
-South	-	37%	40%	-	51%	30%	33%	27%	39%	40%	52%	44%
-West	-	38%	29%	-	24%	44%	48%	47%	41%	31%	34%	24%
Seasonal Migrants	62,000	12,000	33,000	35,000	21,000	37,000	35,000	46,000	56,000	92,000	111,000	84,000
Per of all migrants	27%	10%	15%	18%	11%	18%	20%	18%	20%	20%	29%	21%
Annual tot earnings-$	5,337								1,167	1,294	763	1,088

Source: The Hired Farm Working Force, various years;
*Males only; ** All workers; - means less than 50,000; Seasonal migrants did 75 to 149 days of farmwork

had median earnings of $3,115, of which 72 percent came from 88 days of farmwork. Average earnings were $7,500 for household heads and $3,900, for other migrants, reflecting the high earnings of some migrants. Weighting these average total earnings figures yields average migrant earnings of $5,919 in 1983, or an increase of 137 percent since 1973. Average gross weekly earnings in nonfarm industries, by contrast, rose 93 percent over this time period.

Migrants are eligible for legal services if their incomes are less than 125 percent of the poverty line. In 1983, the poverty line for an unrelated individual was $5,061 and for a family of four $10,178, meaning that the upper income limit for single migrants was $5,061 x 1.25 = $6,326 and $12,723 for families of four. The 1983 HFWF survey reported that 124,000 migratory household heads averaged $7,567 each in farm and nonfarm earnings and that 103,000 migrants who were members of households with migrant or nonmigrant heads averaged $3,937 each (p. 26).

How many migrant families earn more than 125 percent of the poverty line? According to HFWF data, about one-fourth do. The HFWF migrant sample data found 1.72 workers (not all migrants) and 1.95 dependents per migrant household. The HFWF report (p. 34) indicates that 24 percent of migratory household heads had farm and nonfarm earnings of $10,000 or more. Thus, at least one-fourth of all migrants in the HFWF survey appear to have earnings that exceed 125 percent of the poverty line: if the migrants are single, then $10,000 or more is clearly greater than the $6,326 cutoff. If the migrants are families of four, then the $10,000 or more earned by the household head plus 0.72 times (for the partial second worker) the $3,937 average of other migrant workers yields a family income of at least $10,000 plus $2,835 or $12,835, which is greater than the $12,723 poverty line threshold. This $10,000 cutoff is conservative because income for poverty-level data includes additional income such as UI payments, but the HFWF data include only earnings. In a 1983 California survey of migrant and seasonal farmworkers, over two-thirds of the workers reported receiving UI payments during the previous year.

California's UI data reinforce the HFWF finding that almost one-fourth of the migrant households earn more than $10,000 annually. In 1984, the poverty line for an unrelated individual was $5,278 and $10,609 for a family of four, making the legal services assistance cutoffs $6,598 and $13,261. Using the $10,000 cutoff again, that is, assuming that a single worker earning $10,000 or more is ineligible and a household head or dependent earning $10,000 or more is likely to have another household member earning $3,261, then 23 percent of migrant farmworkers in California's UI data have total earnings too high to be eligible for legal services assistance.

The California UI data indicate that 77 percent of the migrants earned less than $10,000 from all farm and nonfarm jobs in 1984, and this UI data permits an analysis of the characteristics of these seasonal "HFWF" migrants (persons with at least two farm jobs in two counties in 1984). Seasonal migrants can be defined in many ways, but one definition is persons whose total earnings were between $1,000 and $12,500 in 1984. In 1984, about 68 percent of the California migrants satisfied this seasonal criterion.

There is no minimum earnings necessary to be considered a migrant farmworker because even one day of qualifying work is sufficient to make a worker a migrant. However, some studies and programs exclude very casual workers; a 1965 legislative profile of California farmworkers excluded all persons with less than $100 in farm earnings, and DOL regulations restrict eligibility for MSFW job training programs to those doing at least 25 days of farmwork. The 1983 HFWF reported that 6 percent of the migrants who headed households or 7,440 persons had less than $1,000 in total earnings in 1983, and 22 percent or 22,660 persons who were members of migratory households earned less than $1,000. (It should be noted that a teenage migrant makes the family a "migratory household.") The California UI data indicate that 14 percent of the migrants earned less than $1,000 from all jobs in 1984; such low-earning migrants averaged 5.5 weeks of work in 1984, including 3.8 weeks of farmwork.

The HFWF data report that about half the nation's migrants also do nonfarm work, suggesting at least one change in occupational title during the year. Between 1973 and 1983, the percentage of migrants who also did nonfarm work rose from 47 to 52 percent. Nonfarm work consistently generates higher daily earnings than farmwork for migrants, the genesis of the "salvage" theory of the farm labor market which holds that the easy-entry farm labor market permits workers to translate otherwise non-wage earning time into wages, albeit at a lower wage than prevails in less accessible nonfarm jobs.

The HFWF does not ask questions about the career intentions of migrant farmworkers, but it does record two kinds of career-relevant data. First, the HFWF records the age of migrant workers. If farm employment is stable, and workers make farmwork a career, then the farm workforce should become older because entry-level workers are not hired. The average age of the farm workforce remains unchanged when employment is stable only if there is a rapid turnover of workers, for example, if new 14 to 24 year-olds replace those leaving the farm workforce.

The migrant workforce stabilized at 200,000 and aged somewhat in the 1970s, according to the HFWF data, but a substantial proportion of migrants continue to be entry-level 14 to 24 year-olds. In 1973, about 53 percent of the workforce was 14 to 24; by 1983, the share of 14 to 24 year-olds had dropped to 35 percent. An increased career attachment may be indicated by the jump in 25 to 54 year-old migrants, from 41 percent of all migrants in 1973 to 59 percent in 1983.

A second indication of migrant career attachment is days of farmwork and the chief activity of migrants during the year. In 1973, 57 percent of all migrants did less than 75 days of farmwork while 25 percent did 150 days or more. In 1983, only 38 percent did less than 75 days and 34 percent did 150 days or more of farmwork. The chief activity of migrant workers was not recorded in 1973, but in 1983 only one-third of the migrants reported that they were primarily not in the labor force because they e.g., attended school.

The distances travelled by migrants were recorded in more recent HFWF reports, and the percentage of migrants who travelled 1,000 miles or more to do farmwork was 29 in 1976, 28 in 1977, 14 in 1979, and 31 in 1983. The number of long-distance migrants interviewed is quite small.

Data from the HFWF sample also provide a contrast to the Lillisand assumptions on migrant workers per household and dependents per migrant household. Lillisand assumed two migrant workers per household in the 19 states "with a heavily Mexican-American influx" (p. 51) and cites USDA as the source for 1.4 workers per household elsewhere. The 1975 HFWF report, on p. 21, says that 67,000 migrants were household heads and 76,000 were other household members, implying only 1.1 migrant workers per migrant household. In the 1983 HFWF migrant sample, there were also 1.1 migrant workers per migrant household.

Lillisand apparently assumed 6.65 dependents per household in the 19 states with a "Mexican-American influx" (p. 52), and noted that this dependent factor was an "average" of the estimates used in three consultant reports and that Lillisand's own survey supported a large migrant family size assumption. As noted before,the definitions of migrant worker and household can be complex: a person can be a migrant worker by doing just one day of migrant farmwork during the year, and the "dependents" in a migrant worker household may include adult and teenage nonfarm workers and nonworking children. The 1983 HFWF migrant sample of 117 migrant households included 131 migrant workers, 164 hired farmworkers, 201 farm and nonfarm workers, and a total 228 nonworking dependents, including 100 young dependents, for the following rations:

-- 1.12 migrant farm workers per migrant household
-- 0.28 nonmigrant farmworkers per migrant household
-- 0.32 nonfarm workers per migrant household in December 1983
-- 1.95 nonworking dependents per migrant household in December 1983
-- 0.85 dependents under age 17 per migrant household in December 1983.

The presence of one migrant farmworker makes the entire household a migrant household in these statistics, so that e.g., an 18 year-old who does migratory farmwork can have a younger brother or sister, a nonworking mother, and a non-migrant father. It is important to note that there are almost two workers per migrant household (1.72), but these workers are 1.12 migrants, 0.28 nonmigrant farmworkers, and 0.32 nonfarm workers.

The HFWF reports include a variety of data on migrant and nonmigrant farmworkers. An analysis of migrant characteristics between 1973 and 1983 indicates that migrants are becoming more minority, older, less educated, and more committed to farmwork. The HFWF has changed the regions to which it distributes migrants several times, but there have been marked changes in the distribution of migrants even across broad census regions. Finally, it appears that the proportion of long-distance migrants bottomed-out during the high gasoline prices of 1979-80; the proportion of migrants who travelled more than 1,000 miles was higher in 1983 than in previous years.

The major demographic changes over the past decade include a drop in the percentage of White migrants from 96 to 45 percent; a rise in the 25 or older group from 47 to 65 percent; and perhaps an increase in the percent of migrants who have eight or less years of education. The average days of farmwork done by migrants was 102 in 1973, fell to a low of 87 in 1979, and then rose to 124 in 1983. Migrants' average total earnings rose 129 percent, and average daily earnings rose 125 percent over the decade. HFWF does not report an average hourly wage: if the HFWF daily wage is divided by eight hours, the implicit hourly wage rose from $2.07 to $4.66 or 125 percent, while the federal minimum wage rose from $1.60 to $3.35 or 109 percent. Finally, the percentage of migrants who only did farmwork varied between 45 and 72 percent, with no trend apparent.

The HFWF distributed migrants across three types of regions between 1973 and 1983. In the four years for which at least partial census region data was reported, from 70 to 85 percent of all migrants were interviewed in the south and west, the reason why the HFWF originally requested a December survey to interview migrants after they had returned home from treks northward, even though the survey results indicate that relatively few migrants travel long distances to do farmwork. Migrants were distributed across 10 standard federal regions in 1977 and 1979, and there was a considerable amount of inter-regional change over this period. For example, Regions 2 (NY and NJ) and 3 (Mid-Atlantic states) jumped from 2 percent in 1979 to 8 percent of all migrants in 1979, Region 6 (the Southwest) dropped from 37 to 24 percent, and Region 8 (Mountain states) dropped from 13 to 7 percent.

Migrants who do 75 to 149 days of farmwork can be analyzed separately. These seasonal migrants were 10 to 27 percent of all migrants between 1973 and 1983, with no trend apparent. There is very little other data reported by HFWF consistently for seasonal migrants.

CALIFORNIA UI DATA

The California UI data defined "HFWF" migrants as persons with at least one crop (01), livestock (02), or selected agricultural service (07*) job outside the county in which the worker had his or her highest earnings from all farm and nonfarm jobs. There were 1,873 such "HFWF" migrants in the 1 percent sample of 9,215 farmworkers in 1984, or 20 percent. This same migrant percentage can be applied to the additional 2,074 sample workers employed only by multi-establishment employers to yield 2,295 sample migrants and 229,500 "HFWF" migrants in California.

California UI data also include information on employers, weeks worked, and earnings, but only for the 1,873 sample "HFWF" migrants with identifiable employers. Among the sample "HFWF" migrants, 1,063 or 57 percent also had nonfarm employers, and 67 percent of those with nonfarm employers had one nonfarm employer, 20 percent had two nonfarm employers, 8 percent had three, 3 percent had four, and 2 percent had five or more nonfarm employers. More "HFWF" migrants had two or more farm employers: only 35 percent had just one farm employer; 24 percent had two; 14 percent had three; 9 percent had four; 10 percent had five or six; and 8 percent had seven or more farm employers in 1984.

Among "HFWF" migrants, there are many workers who do farmwork for only a short time. "HFWF" migrants averaged 12 weeks of farmwork, but 29 weeks of nonfarm work in 1984. Only 57 percent of all "HFWF" migrants had nonfarm jobs, but the migrant sample had more total weeks of nonfarm work than farmwork.

Much of agriculture offers seasonal employment, so any distribution of weeks worked indicates that the small percentage of workers employed almost year-round does a very disproportionate share of the total weeks of farmwork. For example, the 9 percent of all "HFWF" migrants who did 31 or more weeks of farmwork in 1984 contributed 35 percent of all the farm weeks worked. The very casual migrants doing less than 5 weeks of farmwork in 1984 were 45 percent of all migrants but they contributed only 8 percent of all farm weeks worked by the sample "HFWF" migrants.

Seasonal "HFWF" migrants who did 6 to 30 weeks of farmwork were 45 percent of all migrants and they contributed 57 percent of all farm weeks. These 6-to-30 week seasonal migrants earned 52 percent of migrant wages; they had average farm wages of $2,777 in 1984.

The earnings data reinforce other data which suggest that many "HFWF" migrants are really nonfarm workers who work in agriculture outside the urban county in which they obtain most of their earnings. The average "HFWF" migrant had total earnings of $7,575 in 1984, but only $2,400 or about one-third came from farmwork. Weekly wages averaged $263, but were only $203 in farmwork versus $305 in nonfarm work. Hours of work are not reported, but if work weeks averaged 40 hours, then farmwork wages average $5.07 hourly and nonfarm wages $7.62 hourly. The average "HFWF" migrant had 2.8 farm employers and 0.9 nonfarm employers, or the average "HFWF" migrant was employed by four different employers in 1984. Only 15 percent of all "HFWF" migrants had farm earnings in all four quarters of 1984, and only 67 percent had farm earnings in the busy third quarter. Workers employed in the third quarter averaged $1,177, or $392 monthly.

If "seasonal HFWF" migrants are defined as persons who had total earnings of $1,000 to $12,500 in 1984, then two-thirds of all "HFWF" migrants were seasonal and they contributed three-fourths of the farm weeks worked and earned two-thirds of the sample's farm earnings. These "seasonal HFWF" migrants have an employment and earnings profile which conforms to many migrant stereotypes: average total earnings are $4,700, about half of which comes from farmwork; workers work only half the weeks of the year, and half of these weeks worked are farmwork; and 70 percent had farm earnings during the July through September third quarter, but fewer than 45 had farm earnings in all four quarters.

The data on "seasonal HFWF" migrants suggest that most have a substantial commitment to farmwork (they average 13 weeks) but that relatively few find farmwork in all four quarters of the year. Weekly wages are only slightly higher in nonfarm work, suggesting that these seasonal workers could increase their total earnings by doing more weeks of farm or nonfarm work.

The picture painted by UI data suggests that "HFWF" migrants are mostly low-earning nonfarm workers who do farmwork only occasionally during the year. The data for all workers reinforces this "salvage theory" of the farm labor market--across all 12,435 sample workers, the average worker had total earnings of $7,300, of which $2,300 or about one-third came from farmwork. The average sample worker was employed 20 weeks in 1984, but only nine weeks with 01, 02, or 07* employers. For the sample, weekly wages averaged $370, but only $260 in farmwork versus $460 in nonfarm work, or $6.59 and $11.56 hourly for 40 hour weeks. The average worker had 1.2 farm employers and 0.3 nonfarm employers. Only 12 percent of all sample workers had farm earnings during all four quarters; 28 percent had farm earnings in the first quarter, 40 percent in the second, 46 percent in the third, and 32 percent in the fourth.

The salvage theory of the farm labor market asserts that the 80 percent of all "farmworkers" who do farmwork less than year-round are workers who are converting time which would otherwise not generate earnings into paid work time. This theory, propounded by agricultural economists in the 1950s and 1960s, suggested that the best way to help low-earning farmworkers was to reduce the nonfarm unemployment rate, since nonfarm earnings were higher than farm earnings and "farmworkers" presumably wanted to maximize their earnings. The policy implication of this prescription was to draw attention away from low farm wages and farm working conditions and toward impediments to full employment in the nonfarm economy. Farmworker assistance programs relied on this salvage theory of the farm labor market to justify their efforts to get farmworkers out of agriculture, by e.g., training them for nonfarm jobs.

Some 1970s farmworker analyses suggested that the number of farmworkers was declining but asserted that the remaining farmworkers were "professionals" increasingly committed to farmwork. The California UI data does not support this theory of a smaller and more professional farm workforce; instead, it suggests that the farm workforce remains large and diverse. Perhaps the most revealing indication of continued diversity are the similarities between all farmworkers and "HFWF" migrants. In both cases, farm earnings are only about one-third of total earnings; weekly farm earnings are much lower than nonfarm earnings; and only 12 to 14 percent of all workers have farm earnings in all four quarters of the year.

The migrant profile changes considerably if the "averages" are replaced by data for low-earners, middle-earners, and high-earners. The definitions used--total earnings less than $1,000 in 1984, between $1,000 and $12,500, and more than $12,500--are roughly analogous to the casual, seasonal, and year-round terms often applied to farmworkers.

The "HFWF" migrants were 15 percent casual (total earnings less than $1,000), 67 percent seasonal ($1,000 to $12,500), and 17 percent year-round (more than $12,500). Within the seasonal category, most migrants were low-earning seasonal (total earnings of $1,000 to $3,999). However, only 10 percent of these $1,000 to $3,999 migrants had farm earnings in all four quarters of 1984, and they earned an average $2,571 from farmwork.

The "HFWF" migrants were employed by a variety of farm employers. About 72 percent of the "HFWF" sample migrants had at least one job in 1984 with a crop (01) employer, 5 percent had at least one job with a livestock employer, 64 percent had one job with an agricultural services (07*) employer included in the USDA QALS survey; 2 percent had at least one job with other agricultural service employers (e.g, pet and lawn services); and 21 percent had at least one job with a multi-establishment employer.

The average "HFWF" migrant had almost four employers, and very few migrants worked in just one commodity category. For example, 72 percent of the "HFWF" migrants worked on crop farms, but just 7 percent worked only on crop farms. Similarly, almost two-thirds of the migrants worked for at least one QALS agricultural service employer, but only 5 percent worked only for QALS agricultural service employers.

The same gap between the number of migrants with one job in a commodity and migrants who only worked in a commodity occurs for more narrowly defined commodities. For example, almost half of the HFWF sample migrants had one job on a fruit and nut (017) farm, but only 2 percent worked only on fruit and nut farms. Similarly, 265 sample workers or 14 percent had at least one job on a vegetable or melon (0161) farm, but only four workers were employed only on vegetable farms in 1984. The gap between the "at least one job" and "only jobs" in a commodity reflects the fact that 57 percent of the "HFWF" migrants also had nonfarm jobs and the apparent tendency of "HFWF" migrants to shift between crop and agricultural service employers.

SUMMARY

Migrant farmworkers are a heterogeneous subgroup of all persons who do farmwork for wages. The conventional wisdom is that almost all persons who hand-harvest crops are migrants, even though farm labor data have shown consistently that less than 20 percent of all hired workers are migrants. However, the myth that most farmworkers are migrants who follow-the-sun from south-to-north in their families persists; there are such family migrants, but fewer and fewer make the northward trek because they now have other options at home and because northern farm employers prefer easier-to-house solo men, many of whom have been illegal immigrants.

Most migrant studies reinforce the conventional wisdom that migrants travel long distances with their families. However, neither the CPS-HFWF nor the UI data support this stereotype: the CPS-HFWF nor the UI data support this stereotype: the CPS-HFWF data report that most migrants are young white men, although this CPS-HFWF profile is based on data from only 120 to 130 migrants across the United States. The UI data do not generate demographic data, but both UI and HFWF data suggest that about one-quarter of all migrants earn above-poverty-level incomes, fracturing another myth that all migrants are poor. The conflict between statistical and UI administrative data, which picture young whites and often-urban-based migrants and the poor Hispanic family described by migrant assistance programs, suggests that a careful study of migrant characteristics is needed.

Chapter 6

Migrant Workers in Tomorrow's Agriculture

Agriculture is one of America's economic success stories. As a "crown jewel" of the economy, American farmers and farmworkers produce such an abundance of food and fiber that U.S. consumers spend considerably less of their incomes for food than do consumers in other industrial societies. Agricultural abundance has been a product of the U.S. resource base, technology, capital, and labor, and technological breakthroughs promise to continue to make tomorrow's farm problem that of dealing with perennial surpluses rather than food shortages.

Over the past two decades, the U.S. agricultural system has become the subject of criticism. Instead of agreeing that agriculture's success was obvious, critics have argued that a variety of forces are concentrating food production on a relative handful of large farms which obtain government subsidies, misuse pesticides, and remain unwilling to accept nonfarm labor standards for their hired workers. Agricultural spokespersons have often been defensive in the 1980s, arguing that the critics are overemphasizing negative features of a successful food system.

Farm labor is rarely a major issue for most of agriculture; it has been a national issue most recently during Cesar Chavez's 1960s grape boycotts and during the 1980s debate on immigration reform. During these nationwide periods-of-interest in farm labor, it is widely assumed that most farmworkers are poor migrants in need of sympathy and support. Agricultural spokespersons typically respond to these flickers-of-interest in farm labor by emphasizing the diversity in farm wages and working conditions and then arguing that the farm labor market is and has always been the unique provider of jobs for persons who would otherwise be jobless, so that farmers should not be blamed for the misery-at-home and poverty-abroad which generates the pool of people willing to become migrant and seasonal farmworkers.

Farmworker advocates tend to blame farm employers for the plight of the migrant, while farm employers argue that the economy and nature create a seasonal demand for labor that local workers are unwilling or unable to satisfy. Worker advocates argue that there is no shortage of farmworkers, only a shortage of decent wages and working conditions; farmers respond that they have limited control over their prices and thus they cannot afford to pay higher wages and that, in any event, nature ripens crops during the hot summer in an unpredictable fashion, so farmworkers cannot be guaranteed employment. Second-level arguments are usually echos of these: worker advocates urge farmers to diversify, to plant several crops so that workers can be employed for longer periods in one location, and to adopt other farming practices which would enable agriculture to employ fewer workers for longer periods. Most farmers, however, have been reluctant to "decasualize" the harvest labor market because such policies are expensive, requiring the farmer to learn about new crops and markets or requiring the farmer to learn about and spend more to manage hired workers.

There have been changes in the farm labor market which have benefitted workers and employers, but the major theme of most farm labor discussions is how little has changed after a century of debate. The rise of worker advocates, migrant assistance programs, new federal and state laws, and farmworker unions have not extirpated poor migrant farmworkers in the United States. A major effect of this century-of-debate has been to contribute to the harvest-of-confusion over the migrant workforce.

The single most important influence on wages, working conditions, and the structure of the farm labor market has been the number of workers available[1]. This supply of farmworkers, and thus conditions in the farm labor market, will in turn be influenced by the evolving structure of agriculture and by immigration reform.

THE EVOLUTION OF LABOR-INTENSIVE AGRICULTURE

America's founders espoused an agrarian ideology which exalted farming. Thomas Jefferson wrote that "those who till the soil are the chosen people of God," and Jefferson believed that a nation of small family farmers would keep American democracy viable. This agrarian ideology is responsible for still-familiar aphorisms, such as farming is not just a business, it is also a way of life.

What role do labor-intensive farms play in this idealized family farm system? The nation's 2.2 million farms can be sub-divided into a number of categories, such as grouping them by the major commodity that they produce. There are three major commodity subsectors: field crops such as wheat and corn; livestock, dairy, and poultry; and fruits, vegetables, and other horticultural (FVH) specialty crops. Migrant workers are associated with this FVH sector.

The 1982 COA reported that there were 158,000 U.S. farms which produced primarily FVH commodities. Most of these FVH farms are very small and do not employ hired workers; about half reported that they hired workers in the 1982 COA. Even though most FVH farm employers are small, the practice of presenting agricultural averages in which these small entities loom large tends to obscure the concentration in FVH production and thus employment. The 1982 COA, for example, reported that California had over 600 farms which produced lettuce, but the largest producer, the Dole subsidiary of Castle and Cooke, accounts for about one-fourth of California's lettuce production and the ten largest lettuce producers account for about two-thirds of California's and thus U.S. lettuce production. In some commodities, such as California oranges and lemons, the average size of each citrus grove may be small but a dominant marketing cooperative, Sunkist, promotes uniformity in production and labor matters throughout the industry.

Fruits and vegetables are high risk and potentially high profit commodities. An acre is about the size of a football field, and an acre of California strawberries generates an average $25,000 in revenues, versus $150 for an acre of Kansas wheat. The Kansas wheat farmer (and his banker) can pretty much predict that the net return for each acre of wheat will be about $10, because government programs establish price floors and crop insurance provides protection against weather-caused and other losses. The California strawberry producer, by contrast, must invest about $10,000 before the crop is ready to be harvested, and then faces prices that fluctuate between $6 and $12 for a flat of 12 one-pint cartons. The strawberry farmer barely breaks even at the $6 or $7 price, but obtains a very high rate of return at the higher price. Thus, there is a bonanza aspect to FVH farming: producers may scrape by for several years and then enjoy the one year of extraordinary prices and profits necessary to keep them in business another five years.

Fruits and nuts are perennial crops which require planting and then waiting for five to seven years for a crop; vegetables, by contrast, usually mature 40 to 80 days after planting. For a variety of reasons, the vegetable labor market has evolved differently from the fruit labor market, especially in California. The marketers of vegetables tend to exert more control over their production, and most of the marketers are either large public (Dole) or private (Bruce Church) companies or integrated food processors such as Campbells that offer farmers contracts to grow beans or tomatoes for them. The typical California vegetable farm employs 200 or more farmworkers and, if it supplies a fresh vegetable year-round, the company often employs these workers for six to 10 months. These integrated vegetable companies often comprise five or six entities, including a farming company, a trucking company, a marketing company.

The fruit labor market has evolved differently. There are fruit growers who supply fresh fruit to consumers year-round, but the year-round supply of grapes or peaches for U.S. consumers is usually obtained by producing in the United States and by importing from a country such as Chile, meaning that most U.S. fruit workers do not find year-round employment. Farmers must wait five or more years after planting for a return from their fruit trees or vines, and such waits make fruit farming even more speculative in many respects than vegetable farming.

The investment nature of fruits and the frequent practice of having a cooperative or self-owned marketing channel means that the fruit labor market differs from the vegetable labor market. Fruit workers tend to be hired for shorter periods; harvesting e.g., peaches into bags on ladders is considered harder work than picking vegetables; the fruit workforce has fewer women than the vegetable workforce; and there is less opportunity for a fruit worker to migrate with the harvest picking just one crop, as a lettuce worker can do by migrating between northern and southern California.

Both fruit and vegetable farms rely on bilingual intermediaries to find workers. This means that the typical fruit and vegetable grower must communicate with farmworkers through a labor contractor, crew leader, or foreman. These intermediaries are "the boss" in the eyes of most workers, because they often recruit, supervise, and pay workers. Because growers do not communicate directly with workers, each intermediary creates his or her own labor market for 30 to several hundred workers.

The variability in these intermediaries explains much of the persisting controversy over farm labor. Some farm labor intermediaries are honest brokers between growers and workers, but many are not. Since the intermediaries tend to recruit workers who cannot get other jobs, they have a great deal of control over workers. The intermediary may arrange for the worker's transportation, housing, and meals, and charge for each service. The intermediary may illegally undercount a worker's production or make illegal or excess deductions for work equipment. Since the workers are often vulnerable immigrants, few complain.

A central labor feature of fruit and vegetable agriculture is that the labor market has been turned over to such intermediaries. There are a few exceptions, notably at the larger vegetable companies, but most fruit farms that need labor simply turn to a labor contractor or foremen and expect him or her to "round-up" a crew of workers and get the crop picked. This means that hiring has become decentralized in the hands of intermediaries just as production and marketing has become more centralized.

There are several factors which might alter this trajectory of more centralized production and marketing of fruits and vegetables and a more decentralized labor market. The trend toward the concentration of fruit and vegetable production on fewer and larger farms appears to be unstoppable; what may change is how many and what kind of workers are hired in tomorrow's fruit and vegetable agriculture.

Labor-saving machinery has eliminated thousands of FVH jobs. Tree shakers have replaced farmworkers who used to knock nuts such as almonds off of trees, and shakers with canvas catching-frames harvest some of the nation's apples and peaches. Many vegetables such as green beans and processing tomatoes are harvested by machine, and the mechanization of remaining hand-harvested fruits and vegetables has been predicted since the 1960s. However, the availability of an ample and sometimes excessive number of immigrant workers reduced the incentive to develop labor-saving machinery in the 1970s and 1980s, so few major fruit and vegetable harvests have been mechanized since the 1960s.

The major reason why there have been so few recent changes in the way that fruits and vegetables are produced is because immigrant workers have been available. Thus, the future supply of immigrant farmworkers will be the major factor which determines the number and type of farmworkers employed in the United States. If immigrant farmworkers are expensive and hard-to-employ in U.S. agriculture, there will probably be a short-term increase in farm wages and the number of migrants followed by the longer-term development of labor-saving machinery, increased imports of fruits and vegetables, and a slowdown in the growth of consumer demand for hand-picked fruits and vegetables in reaction to higher prices. If immigrant workers are readily available at current farm wages, U.S.-born or domestic migrants will probably continue to be pushed out of farmwork by the preferred immigrants from abroad.

IMMIGRATION REFORM AND MIGRANT FARMWORKERS

The number and characteristics of entry-level farmworkers have been among the most important determinants of wages and working conditions for migrant farmworkers. After the Bracero program which admitted temporary Mexican farmworkers ended in the mid-1960s, farm wages rose, the number of U.S. migrants increased, and farmworker unions became recognized bargaining agents for some farmworkers. As illegal immigration continued in the 1970s and early 1980s, farm wages stabilized or fell and unions lost many of their contracts.

In the mid-1980s, the United States reformed its immigration laws in order to end illegal immigration. The Immigration Reform and Control Act (IRCA) of 1986 should have three major effects on the farm labor market and thus on migrants. First, IRCA imposes sanctions or fines on employers who knowingly hire illegal alien workers. Employers demonstrate that they have not knowingly hired an illegal alien worker by completing an employment verification or I-9 form for each worker hired. This I-9 form requires the worker to have and the employer to see identification which establishes each worker's identity and right to work in the United States, thus deterring illegal alien workers. Sanctions add one more layer of regulation which should diminish the importance of casual farm labor markets in which every one who shows up to work is hired.

IRCA's second effect on migrant workers arises from agriculture's unique legalization program, the Special Agricultural Worker or SAW program. The SAW program permits illegal alien workers who did at least 90 days of qualifying farmwork in the 12 months ending May 1, 1986 to apply for temporary and eventually permanent U.S. legal status. Qualifying farmwork had to be done in Seasonal Agricultural Services, which were defined by commodity (fruits and vegetables of every kind and "other perishable" commodities which range from tobacco to Christmas trees) and activity (doing or supervising fieldwork).

No one knew how many SAW applicants there would be, although most observers put the number of illegal alien farmworkers (mostly migrants) at about 350,000. The number of SAWs soon jumped, however, to over 800,000 in the summer of 1988 and an expected one million by November 30, 1988, when the program ends. Since most illegal alien farmworkers cross county lines and stay away from "home" overnight to do farmwork for wages, most qualify as migrant workers, although some of the farmworker assistance programs are permitted to serve only migrant farmworkers legally-entitled to work in the United States.

The SAW legalization program has confused the migrant worker profile. SAWs are young men from Mexico who apply for legalization in California. The median age of SAW applicants is 28; over 80 percent are male; and almost 80 percent are from Mexico. The SAW application form included space to list ten U.S. farm employers, but in a sample of SAW applicants, about 80 percent obtained their 90 days of qualifying work on one farm doing one task, such as harvesting tomatoes for 92 days on one farm.

If there are 700,000 to 800,000 approved SAWs, and many satisfy the definition of a migrant, does this mean that 70 to 80 percent of the maximum estimate of 1 million migrants are (were) illegal aliens? Not necessarily, because many of the SAW applications are fraudulent, i.e., filed by persons who did not actually do 90 days of qualifying farmwork. Even though SAW applicants could have entered the United States much later than applicants for general legalization (in 1985-86 versus before January 1, 1982), SAW applicants were assumed to have fewer records available to them, so SAWs were permitted to submit affidavits or letters from employers which asserted that they did e.g., 92 days of qualifying work picking tomatoes. Most SAW applicants submitted only such affidavits, not payroll or tax records, and many of these affidavits were sold by labor contractors and others who did not employ the SAW applicant as stated in the letter.

Some SAW fraud is transparent, as when SAW applicants asserted that they picked strawberries from ladders or picked baked beans to qualify for legalization. However, most of the SAW applicants make reasonable-sounding claims, such as a SAW asserting that he harvested broccoli for 92 days. There is no easy way to detect such fraud on an individual basis, but an analysis of data on workers reported to California Unemployment Insurance authorities suggests that there must be enormous employer under-reporting to UI authorities, massive fraud among SAW applicants, or some of both.

In 1985, California employers who produced the commodities which qualify as Seasonal Agricultural Services (SAS) reported a total 806,000 workers (SSNs) to UI authorities; these workers include fieldworkers and their supervisors as well as persons with nonfarm occupations, such as clerks, mechanics, and salespeople. Most of these SAS workers will not qualify for SAW status because they did not do enough work on SAS farms; in 1985, 350,000 earned less than $1,000 from SAS employers, and it is very unlikely that many of these workers would have averaged just $11 daily over 90 days. At the other end of the SAS earnings spectrum, few SAS fieldworkers are likely to earn more than $12,500 annually; hence, qualifying SAW workers should be among the subset of 370,000 workers whose 1985 SAS earnings were between $1,000 and $12,500.

There are two ways to convert the 90 days of SAS work into the workers reported to California UI authorities. One conversion is from the SAW 90-day work requirement to UI data on weeks worked for SAS employers. Since IRCA requires at least 90 days of qualifying work, the UI data can isolate persons who had at least 18 weeks of work with an SAS employer in 1985 (18-5 day weeks is 90 days). However, to avoid the inclusion of farm managers, clerks, and other non-field workers, an upper limit on weeks worked must be established. USDA defines persons employed 150 or more days on one farm as regular or year-round workers; the calculations below used 40 weeks as the upper limit because few fieldworkers are likely to find employment 5 days each week over extended periods (most SAW applicants reportedly need at least 25 weeks to obtain 90 days of SAS employment).

The workers employed 18 to 40 weeks by SAS employers are one universe from which California SAW applicants can be drawn. About 115,000 California workers were employed in SAS commodities for 18- to 40-weeks in 1985; they averaged $6,100 for 27 weeks of SAS work. Most of these 18- to 40-week SAS workers satisfy the seasonal farmworker stereotype: three-quarters earned $1,000 to $7,500 and this subgroup averaged $4,400 for 25 weeks of SAS work. Only half of the 18-to 40-week SAS workers had more than one farm employer in 1985, suggesting that many SAW applicants will need to list just one employer to satisfy the 90-day work requirement. About two-thirds of these 115,000 SAS workers satisfy the definition of seasonal workers, but only one-quarter satisfy the migrant definition[2].

Alternatively, the universe of SAW applicants can be approximated by isolating the workers who had qualifying earnings from SAS employers in 1985. It is hard to translate SAS earnings into days of farm work, especially because a day of farm work for the SAW program is defined as one hour or more. However, the SAW program permits applicants to estimate their days worked on the basis of earnings, e.g. some SAW applications say that the estimated 100 days of qualifying work was based on 1985 earnings of $4,000, an hourly wage of $5, and an average 8 hours of work per day.

Farmworkers are paid hourly and piecerate wages. Hourly wages in California average $4 to $5, and piecerate earnings are $5 to $6 per hour. Hourly workers usually work 8 or more hours daily, while piecerate workers average 6 to 7 hours of work. Assuming a minimum $40 daily wage, SAS workers would have to earn at least $2,700 from SAS employers to qualify for SAW status. A rough approximation of the SAW universe based on earnings data might include one-half of the workers who earned $1,000-to-$3,999 from SAS employers in 1985 (93,840), all of the $4,000-to-$7,499 (72,580), and one-half of the $7,500-to-$12,500 group (21,280), or a total of 187,700 potentially SAW-eligible workers in California.

This analysis of the data reported by SAS employers to UI authorities in 1985 indicates that California's potential SAW universe ranges from 115,000 to 188,000 workers. Of course, not all of these SAW-eligible workers were illegal aliens: a September 1987 survey of farm employers found that employers believed 42 percent of their seasonal workers were illegal aliens who would apply for the SAW program. Applying this percentage to the UI data yields 48,000 to 78,000 SAW-eligible workers.

These UI data indicate that the SAW programs has been too successful in California and perhaps elsewhere. As of September 1988, about 435,000 SAW applications had been filed in California, 54 percent of the 800,000 SAW applications nationwide. The INS completed its reviews of 267,000 SAW applications by September 1988, and approved 88 percent of them; INS has said that this approval rate will probably fall as more fraudulent applications are identified, e.g., the INS suspects fraud in half of its "open" SAW cases. However, even an 80 percent approval rate applied to 500,000 SAW applications in California yields 400,000 SAWs, more than most observers anticipated and more than would be expected if less than half of the 188,000 workers in the UI data were illegal aliens. The number of SAWs is also surprisingly high because almost 50,000 illegal alien farmworkers became legal U.S. residents under the general legalization program which ended in May 1988.

The UI data conflict with the SAW data, that is, either California farm employers reported only one-third of their 1985 employees to UI authorities, or two-thirds of the SAW applications are fraudulent; there may also be a combination of under-reporting and fraud. It should be emphasized that this conclusion is only suggestive because the UI data may underestimate the SAW-eligible universe if workers used several Social Security Numbers (SSNs) in 1985 to accumulate SAS weeks and earnings, so that the earnings cut-off is too strict; in one sample, about 20 percent of the SAW applicants had more than one SSN in 1985-86. The SAW-eligible universe may also be larger than suggested by the UI data if SAW applications filed in California included qualifying work done in other states such as Oregon and Washington, although only 2 percent of the California SAW applications in one sample included any work done outside of California.

SAW fraud confuses the migrant profile, and the third element of IRCA, the H-2A and Replenishment Agricultural Worker (RAW) programs, may confuse the migrant picture even more in future years. IRCA assumed that sanctions would end illegal immigration, the SAW program would legalize illegal alien farmworkers, and then, if "supplemental" foreign farmworkers were still needed, they would enter the United States through either the H-2A or the RAW program. Since H-2A's and RAWs would be migrants, both would alter the U.S. migrant worker profile.

The H-2A program is a contractual foreign worker program which requires farmers to plan their employment needs, advertise for American workers and offer them at least a minimum DOL-approved wage and housing package, and then, if American workers are not available, farmers are "certified" to go abroad and recruit foreign workers. These H-2A foreign workers then receive temporary visas to enter the United States to do farmwork. H-2A workers are migrants by definition, since they leave their normal foreign residence and, by coming to the United States to do farmwork for wages, they cross state and county boundaries.

The RAW program is an alternative greencard immigrant program. Instead of farmers asking DOL to certify their individual needs for supplemental foreign workers, the RAW program permits a certain number of immigrants to enter the United States; these RAW workers can remain in the United States indefinitely so long as they fulfill the terms of their visas, which require, inter alia, at least 90 days of SAS work annually for three years. This means that RAW workers will enter the United States, seek farmwork at prevailing wages on SAS farms, and then, after doing at least 90 days of farmwork, the RAWs can switch to nonfarm work.

Both the H-2A and RAW programs provide controversial additions to the migrant workforce. The H-2A program is controversial because it provokes litigation over the minimum wage and housing package which employers must offer in their bid to obtain American workers. Clearly, at some high-enough wage, Americans would migrate to do farmwork and there would be no need for H-2A workers; at these higher wages, farmers may also not hire as many workers, and thus be able to find all the workers they "need" within the U.S. The H-2A program has contradictory goals: it is impossible to satisfy the conflicting goals of protecting American workers from "unfair" competition from H-2A migrants; keeping most American farmers in business and having their fruits and vegetables harvested by hand; and providing American consumers with U.S.-produced food. These conflicting goals have made the H-2A program among the most litigious in the United States.

The RAW program may generate a different kind of controversy. The RAW program cannot admit workers before October 1, 1989, so much of the controversy surrounding it is hypothetical.

The RAW program requires two separate calculations. First, a *ceiling calculation* based on the number of SAWs is made to

establish the maximum number of RAWs who can be admitted in
FY 1989. Then a second *shortage calculation* is made to determine
whether there will be a shortage of labor in SAS in FY 1989. The
smaller of these two numbers controls RAW admissions, for
example, if the ceiling calculation yields up to 160,000 RAWs and
the shortage calculation yields 100,000, only 100,000 RAWs will
be admitted.

RAW Ceiling. The maximum number of RAWs in FY 1989 is
95 percent of the number of SAWs, minus the number of SAWs
who did at least 15 days of work in SAS in FY 1988, and plus or
minus the change in the number of H-2A workers in SAS crops
admitted in FY 1989 versus FY 1988.

For example, if the number of approved SAWs is 800,000, then
the ceiling on RAW admissions in FY 1990 is 760,000 minus say,
600,000 SAWs who did at least 15 days of SAS work in
FY 1988, and minus say 10,000 additional H-2A workers if H-2A
admissions increase from 20,000 in FY 1988 to 30,000 in
FY 1989. The RAW ceiling for FY 1989, in this example, is
150,000.

The ceiling calculation is made in terms of people, while the
shortage calculation is made in man-days worked. This means that
the independent ceiling and shortage calculations may conflict. The
ceiling calculation, for example, might include 600,000 SAWs who
average as few as 20 or as many as 80 days of SAS work; the
higher average days worked represents, in this case, four times
more man-days of farmwork from SAW workers.

Shortage Calculation. The RAW ceiling calculation establishes
only the maximum number of RAWs who can be admitted; none can
actually be admitted unless USDA and DOL agree that there will be a
labor shortage in SAS agriculture in FY 1989. Shortage
calculations are made in man-days and include the work done or
likely to be done by a variety of workers, including SAWs,
domestic migrant and seasonal workers, year-round hired hands on
certain farms, and even a paid teenager working on the family's
Iowa grain farm. The calculations involved to determine the gap
between needed SAS man-days in FY 1989 and available man-days
in FY 1989 are complex, and they will be based on small sample
estimates which make heroic assumptions.

To determine whether there will be a labor shortage in SAS ·
agriculture after October 1, 1989, USDA will estimate the demand
or need for labor in SAS; DOL will determine the supply or
availability of labor to SAS; and farm employers will report how

many SAW workers they employed and for how long. If, for example, USDA determines that there were 180 million man-days worked in SAS in FY 1989 and that there are no changes expected in this demand or need number in FY 1990, and if DOL determines that about 20 percent of the SAS man-days are lost annually because of exiting workers, and that no new workers will be available to SAS, then the shortage number is 20 percent of 180 million or 36 million man-days.

This man-day shortage number must be converted into RAW visas or workers. The Bureau of the Census will analyze employer reports to determine the average number of SAS days worked by SAWs in FY 1989 and, if this average is 90 days, then the estimated shortage is 36 million man-days divided by 90 or 400,000 RAWs. In this example, only 150,000 RAWs could be admitted even though the shortage was 400,000 because the RAW ceiling calculation yielded only 150,000 RAWs.

Regardless of which criterion ultimately governs RAW admissions, there may be a great deal of confusion over the activities of RAWs in the United States. For example, if RAW workers are admitted and they flock to California because it has the highest minimum wage for farmworkers, ($4.25 per hour in July 1988 versus $3.35 elsewhere), will Washington farmers complain about localized labor shortages and obtain the emergency admission of more RAWs, who may simply migrate to California because RAWs are "free agents" in the U.S. labor market? As long as there are wage and working condition differences between farm labor markets, migrants (including RAWs) will move between them, generating complaints from farmers about insufficient workers. The alternative to admitting supplemental foreign workers is, of course, to encourage farmers without enough workers to raise wages and benefits, mechanize, or change crops.

SUMMARY

Migrants are one of the many paradoxes which riddle U.S. agriculture. In the U.S., family farms have been revered as efficient producers of food for a hungry world, while migrant workers have been pitied for being among the most ill-housed, ill-fed, and ill-paid Americans. Migrants are associated with fruits and vegetables, which have become more important in the American diet as health-conscious and affluent Americans consume more of them. This means that the eating habits of Americans perpetuate the very farm labor market conditions that many Americans would like to extirpate.

Immigration reform has created a harvest-of-confusion in the farm labor market. According to SAW legalization data, there were many more illegal alien farmworkers than most employers, assistance organizations, and researchers suspected. However, SAW legalization may be painting a misleading picture of farmworkers because many of the SAW applications appear to be fraudulent; according to the California UI data, at least half are fraudulent. Thus immigration reform, which was expected to generate a tabula rasa from which a more accurate understanding of the farm labor market could proceed, instead has created even more confusion about how many and what kind of persons work in U.S. farms.

The number and characteristics of farmworkers is the single most important determinant of whether the plight-of-the-migrant will still be a nagging blotch on U.S. agriculture in the 21st century. If immigrant workers are readily available to U.S. farmers, then the farm labor market will not adjust and become more like nonfarm labor markets, instead, it will become more and more isolated, offering jobs that Americans refuse. The alternative to an isolated and immigrant-dominated farm labor market is the brave new world of wages and working conditions sufficient to attract American workers to farms; such a world might make "migrant farmworker" no more of a reason for pity than travelling salesperson.

NOTES

1. Farm labor structure refers to features such as who actually hires worker (foremen, labor contractors, or farm operators); what wage system is used to pay them (hourly, piecerate, or salary); and whether farm employers have a mechanism for identifying and encouraging the best workers to remain or return.

2. In the UI analysis, a seasonal worker had farm (01, 02, 07*) earnings of $1,000 to $12,500 and did 5 to 30 weeks of farmwork. A migrant worker had at least two farm jobs in two counties.

SELECTED BIBLIOGRAPHY

Barnett, Paul. Assessment of Needs of the Rural Poor in California: A Research Report (Salinas, CRLA, 1986). This report reviews California farm labor force data, the role of FLCs, mid-1980s farm production and marketing issues, and working conditions in rural California. A 10-page appendix discusses farm labor data.

Booth, Philip. "Coverage of Agricultural Workers," Unemployment Compensation: Studies and Research (Washington: National Commission on Unemployment Compensation, 1980), Vol. 3, pp. 673-704. A valuable review of the debate over providing UI to seasonal farmworkers.

Brown, G.K. "Fruit and Vegetable Mechanization," in P. Martin (ed) Migrant Labor in Agriculture: An International Comparison (Berkeley: Giannini Foundation, 1984) pages 195-209. In the early 1980's there were about 7 million acres of fruits and vegetables grown in the U.S., and they utilized an average 120 hours per acre in production and harvesting. The major labor-intensive fruits are citrus(1.3 million acres or about 29 percent of U.S. fruit acreage), grapes (17 percent), apples (13 percent), peaches (6 percent), plums and prunes (3 percent) and pears (2 percent). The major vegetables are potatoes (1.2 million acres or 38 percent of vegetable acreage), sweet corn (18 percent), tomatoes (13 percent), green peas (10 percent) and snap bean and lettuce (8 percent each).

There is much more mechanical harvesting of vegetables than of fruits. Lettuce is the only major acreage vegetable which is completely hand-harvested but almost all of the citrus, apples, peaches and pears are hand-harvested.

Cargill, B. F. and G. E. Rossmiller. Fruit and Vegetable Harvest Mechanization (East Lansing; Rural Manpower Center, 1970). Three reports on the technological, manpower, and policy implications of fruit and vegetable mechanization. In 1968, there were an estimated 1.1 million jobs on fruit and vegetable farms, a number that was expected to shrink to 900,000 by 1975 despite increased production because of bean, cucumber, and grape mechanization . The authors of this report expected rapid labor-saving mechanization that would perpetuate "an overall surplus of farm labor" (p. 17) and thus require a variety of education and training programs to assist farmworkers who "will be moving from farm to nonfarm work" (p. 22). The fruit and vegetable workforce "will consist of a core of skilled and semi-skilled men operating farm machinery supplemented seasonally by unskilled manual laborers, mostly local housewives and school-age youths with a small component of interstate migrant laborers," p. 18).

Centaur Associates, Baseline Analysis and Economic Impacts of Proposed Field Sanitation Standard. Centaur Associates for OSHA, April 10, 1984. Estimates that 565,000 farms employ field workers who do hand labor and that 52,000 of these hire 11 or more workers (p. 12). These 52,000 large farm employers are estimated to employ 529,000 of an estimated 2,492,000 hired workers, and these 529,000 workers average 4.7 months of farmwork each. Vegetables account for 38 percent of the affected farmwork, fruits 28 percent, and tobacco 23 percent (p. 13). This study makes heroic assumptions: it asserts that, e.g., the minimum-sized citrus or grape farm that hires 11 or more workers is 100 acres (p. 20); estimates the percentage of total citrus production on such "large" farms; and then multiplies the percentage of production on large farms by the 1981 HFWF distribution of workers among crops to determine the number of affected workers. This study is based on Census of Agriculture, HFWF, and "case study" data.

Coltrane, Robert. Immigration Reform and Agricultural Labor (Washington: USDA, ERS-AER 510, 1984). This report uses 1978 Census of Agriculture wage data to determine that two states--California and Florida--contained 40 percent of the fruit, vegetable, and horticultural (FVH) farms and that the 10 states with the most such farms paid 81 percent of total farm wages. Wages averaged 13 percent of reported total farm production

expenses across agriculture, but labor as a percentage of
production costs ranged from 5 percent in beef cattle to
56 percent in FVH agriculture. All types of agriculture employ
mostly seasonal workers, but the ratio of seasonal workers
(employed less than 150 days) to year-round workers ranges
from 8:2 on FVH farms to 2:1 on poultry farms.

Craig, Richard B. The Bracero Program: Interest Groups and
    Foreign Policy (Austin: University of Texas Press, 1971). An
    excellent treatment of the politics of the Bracero Program.
Daberkow, Stan and Leslie Whitener. Agricultural Labor Data
    Sources: An Update (Washington: USDA, ERS-Agricultural
    Handbook, 658, 1986). This is the best recent study of farm
    labor data sources. It includes references to source data and
    several figures which compare data sources; it does not include
    recent data from these sources.
Daniel, Cletus E. Bitter Harvest: A History of California
    Farmworkers 1870-1941 (Berkeley: University of California
    Press, 1981).
Dement, Edward. Out of sight, Out of Mind: An Update on
    Migrant Farmworker Issues in Today's Agricultural Labor
    Market, Working Paper prepared for the National Governors'
    Conference, March 1985. The Governors' assume that the
    federal government has a special interest and responsibility for
    migrants, Indians, and refugees in education, health care, and
    employment. Dement concludes that (1) there is disagreement
    about the number of migrants, (2) that no data gathering system
    is accurately counting migrants, and (3) that federal and state
    governments should end the "special exclusions" of
    farmworkers under labor laws, coordinate assistance efforts,
    and focus on workers in fruit and vegetable agriculture.
Emerson, Robert (Editor). Seasonal Agricultural Labor Markets in
    the U.S. (Ames, Iowa: Iowa State University, 1984). A
    collection of 14 papers prepared for a DOL-sponsored
    conference held in 1981. The chapter by Emerson deals with a
    sample of 250 migrant men in Florida who in 1970-71 averaged
    six years of education, 16 years experience, worked 45 weeks,
    and were 86 percent non-White. Emerson's analysis suggests
    that migration was a rational response to (un)employment and
    wage differences in the early 1970s, especially for Black Florida
    families.
        Varden Fuller's introduction defines the "functioning" of
    seasonal labor markets in terms of the "problems" associated

with them: migrancy, alien labor, low earnings, etc. Fuller noted that the major change in seasonal farm labor over the past 50 years was the decline in "migratoriness". According to Fuller, migrancy gave "absolution" for farm employers and local officials because everyone could assume that workers would show up when needed and then disappear. Fuller concludes that the seasonal farm labor market may be "a bit more stable and less uncertain" after 50 years and that the "lack of labor market structure may not be all bad" as society gropes for the proper trade-off between public assistance and work for persons who are not regularly in the workforce.

Fisher, Lloyd. The Harvest Labor Market in California. (Cambridge, Ma: Harvard University Press, 1952) This is one of the classic studies of the farm labor market.

Galarza, Ernesto. Farmworkers and Agri-business in California, 1947-1960 (Notre Dame: University of Notre Dame Press, 1977). This book recounts Galarza's fortune's as a labor organizer during the Bracero program. Galarza wrote that between 1952 and 1959, when the Bracero program was at its peak, U.S. farmworkers faced "relocation under the sustained pressure of the bracero system." (p. 242).

Holt, James et. al. Toward the Definition and Measurement of Farm Employment (Washington: USDA and AAEA Workshop, May 1977). This report is a comprehensive review of farm labor data sources which emphasizes the difficulties inherent in obtaining reliable data, viz, large number and dispersion of farm employers; seasonal fluctuations in farmworker employment; the large number of short-term or casual workers who may also have other jobs; and the lack of standard definitions and a central agency responsible for farm employment data. The report reviews major household and establishment data sources and recommends that farm employment concepts be made as comparable as possible to nonfarm concepts.

Jamieson, Stuart. Labor Unionism in American Agriculture (Washington: Bureau of Labor Statistics Bulletin 836, 1945). The most comprehensive treatment of farmworker unions through 1945.

Jenkins, J. Craig. The Politics of Insurgency: The Farm Worker Movement in the 1960s (new York: Columbia University Press, 1985). This book outlines sociological theories of poor peoples' movements, the characteristics of California agriculture, and the characteristics of three post-WWII farmworker unions--the

NFLU, the AWOC, and the UFW. The book emphasizes the changed political climate which helped the UFW succeed in its organizing and boycotts.

Kestenbaum, Bert. "Social Security Farmworker Statistics, 1977", Social Security Administration Note No. 3, May 14, 1981. Reports 2,443,000 farmworkers in 1977 employed by 350,000 farm employers who paid $6.27 billion in wages; 81.1 percent of the farmworkers had only one farm employer.

Larson, Alice. Data Source Review and Comparative Estimates for Migrant Workers in Washington State, 1986. This report compares the migrant worker estimates from various sources, e.g., the sum of the peak monthly ES-223 estimates for Washington was 30,638 between 1981 and 1984; HCR estimated a peak 109,486 for the early 1980s; and 12,553 peak migrant children were enrolled in the MSRTS in 1984-85. The report concludes that the 1983 Migrant Health estimates "be considered accurate given the present state of information available." The Migrant Health estimates take peak month ES-223 data, increase it for dependents, adds 29 percent to correct for "undercounting", and the resulting range is a migrant worker and dependent population of 44,198 to 139,247 for Washington.

Lillisand, Davis, Linda Kravitz, and Joan McClellan. An Estimate of the Number of Migrant and Seasonal Farmworkers in the U.S. and the Commonwealth of Puerto Rico (Washington: Legal Services Corporation, May 1977). This report includes a critique of farm labor data, estimates of the populations of migrant and seasonal workers and dependents for each state in 1976, a critique of migrant characteristics, and an extensive bibliography.

Littlefield, Carla and Charles Stout. Colorado Migrant Farmworker Health Survey: Preliminary Report, March 30, 1987. This study attempted to estimate samples of the: (1) families and (2) solo males aged 18 to 50 in four areas of Colorado between July 1, 1986 and September 30, 1986: north central, southeast, south central, and western slope. Migrants were defined as persons who established a temporary residence in the past 24 months for the purpose of engaging in seasonal agricultural labor. The sampling frame was a list of temporary residences "usually occupied by migrants" in each area (p. 10); this means that migrants could be interviewed only if they were in a temporary residence between July and September 1986.

Separate lists were constructed for family housing units and dormitory beds. Of the 1,386 units and beds, 598 were targeted for interviews and interviews were completed with 331 or 55 percent of the target population. The critical factor in this survey design is whether the residence list was comprehensive enough to include all potential migrants, such as the sons and daughters of farm operators who stayed overnight with a relative to do farmwork, or whether the sample included only those migrants likely to satisfy the Hispanic stereotype.

The sample--94 percent Hispanic--was 129 family males, 126 family females, and 74 solo males. Solo males had a mean age of 29, versus 32 for family workers. Most interviewed migrants had "permanent residences" in Texas (51 percent) or Mexico (28 percent). The sample averaged six years of education, 74 percent were married, and 77 percent spoke mostly Spanish. At the time of the survey, 42 percent were working full-time (not defined) and 49 percent part-time. The mean individual income (for 1985) was $3,600; the median $3,000; for families the mean was $6,400 and the median $6,000. About 10 percent of family units interviewed had annual incomes of $11,000 or more. The average number of working family members at the home base was 2.9; there were an average 6.2 total family members, implying 3.3 dependents per family or 1.1 dependents per migrant worker.

Martin, Philip. "California Farmworkers: The 1984 UI Data," Agricultural History Center Working Paper, UC Davis, 1987. This paper analyzes a one-percent sample of the employment patterns of persons or social security numbers reported at least once by a California farm employer in 1984.

Martin, Philip, Suzanne Vaupel, and Daniel Egan. Unfulfilled Promise: Collective Bargaining in California Agriculture. (Boulder: Westview Press, 1988).

Mamer, G.W., "The Use of Foreign Labor for Seasonal Farm Work in the U.S. : Issues Involved and Interest Groups in Conflict" Journal of Farm Economics Vol. 43, December 1961 pages 1204-10. This article reviews the effects of the Bracero program on farm labor markets: Lower wages and less uncertainty about the supply of seasonal workers for employers; and depressed wages and working conditions and displacement for American farmworkers. This review notes that there are few quantitative studies of the Bracero program's impacts.

McWilliams, Carey, <u>Factories in the Field: The Story of Migratory Farm Labor in California</u> (Santa Barbara: Peregrine Publisher, 1939 and 1971). This classic "social history" of California agriculture was published in 1939, the same year that John Steinbeck's the <u>Grapes of Wrath</u> was published. The major theme of McWilliams book is that the large land parcels assembled by land speculators and railroads and the availability of cheap immigrant labor were mutually reinforcing factors which brought about factories in the fields. Steinbeck's book recounts the biblical-style drama of how Dust Bowl refugees who had moved across to desert to the promised land confronted landowners who had grown accustomed to docile immigrant workers. The two books generated an enormous amount of publicity--the Associated Farmers called McWilliams "Agricultural Pest No. 1"--and the controversy created by them was largely responsible for bringing the United States Senate's La Follette Committee to California.

<u>Migrant and Seasonal Farmworkers in Texas</u> (Governor's Office of Migrant Affairs, Austin, July 1976). This was the first of what was intended to be an annual survey of farmworkers in Texas. The sampling procedure was to obtain lists of farmworker households from service agencies and then develop a counted-to-uncounted ratio of farmworkers. The report concluded that there were 85,600 MSFW households in Texas which included 495,000 persons, or 5.8 persons each. About 95 percent of the MSFWs were Hispanic; half received assistance from at least one MSFW program; and over one-third of the migrants were intra-Texas migrants.

<u>Migrant and Seasonal Farmworker Programs</u> (Washington: DOL-ETA, 1978). In 1973, CETA mandated DOL Title 3, Section 303 to deal with the employment problems of Migrant and Seasonal Farmworkers (MSFWs). The booklet begins with grim statistics about farmworkers: average years of education are 8.5; 9 workdays of 9 to 12 hours; 60 percent of families earn less than $6,000; life expectancy is 20 years less than average; and fewer than 10 percent receive public assistance despite their obvious need for it.

In the mid-1970s, DOL reported that MSFWs were mostly in California, Texas, North Carolina, Florida, Wisconsin, Puerto Rico, Minnesota, Iowa, Washington, and Pennsylvania. These state rankings reflect DOL's assistance allocations, which were based for FY 1975 on ES-223 data and for 1976 and 1977 on WALS data. MSFWs eligible for 303 assistance had to have or be a member of a family which has total earnings below the poverty line or be receiving cash welfare assistance and have earned more than 50 percent of their income from farmwork during 12 of the previous 18 months. In addition, a seasonal farmworker had to do at least 25 days of work during the previous year in crop or livestock production but not more than 5 months for any one employer; supervisors and full-time students are excluded. Migrants are the subset of seasonals unable to return home at the end of the workday.

In 1976, most of the MSFW participants in OJJ, classroom, and work experience programs were economically-disadvantaged Blacks. DOL programs enroll or affect MSFWs only for a short time; about 90 percent of the 1976 enrollees received "supportive services" such as information or counseling.

This booklet also outlines the history of MSFW employment and training programs. Title 111-B of the Economic Opportunity Act of 1964 included $40 million for farmworker programs; these were transferred to DOL in 1973. The Secretary of Labor in June 1971 created the National Migrant Worker Program, which spent about $26 million over 4 years to help Employment Service agencies and others to provide training and placement services to migrants. Section 303 of CETA in 1973 was the first major categorical MSFW employment and training program administered at the national level; it spent $63 million in FY 1976. In 1976, farmworker assistance funding was about $233 million: USDA housing $9 million: Migrant Education $123 million; Migrant Health $28 million; Migrant day care $10 million; and 303 programs $63 million.

Migratory Labor in American Agriculture (Washington: Presidential Commission on Migratory Labor, 1951). The Commission report defined a migrant as a farmworker who did not "maintain a stable home the year-round," and notes that "it is only in agriculture that migratory labor has become a problem of such proportions and complexity as to call for repeated investigations by public bodies." (p. 1). The Commission estimated that there were 1 million migrants in 1950: 500,000 domestic migrants; 100,000 legal braceros and H-2 workers; and 400,000 illegal Mexican workers. Statements on the 1950s prospects for mechanization are instructive: "We have in the research laboratories machines which will take care of much of the need of additional labor in the citrus industry. Given another three or four years . . . I don't think the citrus industry will be in need of this migrant labor" (p. 15). "[Cotton-harvesting] machines will never replace all of the labor" (p. 16). In the mid-1980s, virtually all cotton is mechanically-harvested and all citrus is hand-picked.

Mines, Richard. and P. Martin. A Profile of California Farmworkers (Berkeley: Giannini Foundation Information Series 86-2,1986). This report analyzes survey data from 1286 farmworkers in California in 1983.

Moore, Truman. The Slaves we Rent (New York: Random House, 1965). This book is the source of the oft-made statement that the U.S. has better data on migratory workers than it has on migrant farmworkers.

Morgan, L.C. and B.L. Gardner, "Potential for a U.S. Guestworker Program in Agriculture: Lessons from the Braceros", pages 361-411, B. Chisuick (ed) The Gatewaty: U.S. Immigration Issues and Policies (Washington: AEI, 1982). This paper estimates the supply of and demand for farm labor in the 7 states with 90 percent of all Bracero labor: Arizona, Arkansas, California, Colorado, Michigan, New Mexico, and Texas. The model indicates that the Bracero program on average reduced farm wages by about 9 percent, that is, the average (1953-64 average) wage rate, in 1977 dollars, would have been $1.58 instead of $1.44 if there had been no Bracero program.

The model makes a state's demand for farm labor (in thousands of hired workers) a function of : (1) the hourly 1977 farm wage; (2) labor's share of production expenses; (3) the prices received by farmers the previous year; (4) the prices paid by farmers, (5) acres of cropland harvested; and (6) agricultural productivity. The state's employment of hired farmworkers is a function of: (1) the hourly 1977 farm wage; (2) the 1977 hourly manufacturing wage; (3) the unemployment rate of UI-covered workers; and (4) the number of Braceros. The number of Braceros is a function of: (1) the 1977 Mexican minimum farm wage; (2) the hourly 1977 (U.S.) farm wage; (3) a dummy variable for the years 1953-64; and (4) another dummy variable for years after 1961.

Nazario, Sonia. "More Farmworkers Are Finding New Jobs, Settling in One Place," Wall Street Journal, June 26, 1986, p. 1. A newspaper report on "2 million" U.S. migrants, and

especially the "150,000" in Florida who are increasingly pushed
and pulled out of the migrant stream. Nazario reports that
80 percent of Florida's "150,000" migrants are Hispanic, and
that half of these Hispanic migrants are illegal aliens (sources for
these statistics are not discussed). Migrants are being pushed
out of migrancy by competition from illegal alien workers and
lower real wages, mechanization, and continued problems with
housing and child labor. The report profiles several migrant
families who have been pulled into nonfarm jobs by the growth
of service jobs in rural Florida, migrant parents' desires to
educate their children, and migrant assistance programs.

Oliveira, Victor and Jane Cox. The Agricultural Work Force of
1985 (Washington: U.S. Department of Agriculture, Economic
Research Service, Agricultural Economics Report 582, 1988).

People Who Follow the Crops, U.S. Commission on Civil Rights,
1978. The report begins with this sentence: "The migrant
farmworker is, by definition, a poor person."

Pollack, Susan L. The Hired Farm Working Force of 1979
(Washington: USDA, ERS-AER 473, 1981). Reports that
2.7 million persons 14 and older did farmwork for wages in
1979. Farmworkers averaged $2,444 for doing 102 days of
farmwork. Migrants (217,000) were 63 percent White and
27 percent Hispanic and averaged $2,277 for 87 days of
farmwork.

Pollack, Susan L. The Hired Farm Working Force of 1981
(Washington: USDA, ERS-AER 507, 1983). Reports that
2.5 million persons 14 and older did farmwork for wages in
1981. Farmworkers averaged $2,6549 for 98 days of
farmwork. Migrants (115,000) were 75 percent White and
17 percent Hispanic and averaged $2,728 for 112 days of
farmwork.

Pollack, Susan L. The Hired Farm Working Force of 1983
(Washington: USDA, ERS-AER 554, 1986). Reports that
2.6 million persons 14 and older did farmwork for wages in
1981. Farmworkers averaged $3,138 for 100 days of
farmwork. Migrants (226,000) were 45 percent White and
15 percent Hispanic and averaged $4,638 for 124 days of
farmwork.

Pollack S., R. Coltrane, and W. Jackson. Farm Labor Wage Issues
(Washington: USDA, ERS-EDD Report 820615, 1982). This
report states (1) that the substitution of the minimum wage for
the AEWR would have lowered the wage bills of farms using H-

2 apple, tobacco, and sugarcane workers by about 19 percent, and (2) covering all farmworkers under the federal minimum wage (not just those employed on farms that had 500 or more mandays of labor in one quarter of the previous year) would have increased 1980 farm wages less than 4 percent. Almost 90 percent of the difference between the AEWR and minimum wage occurs in Florida sugarcane, where wages would have been reduced from $4.09 to $3.10 in 1980. Some of the data used to estimate employer costs is dubious: despite charging H-2 workers $4 to $5 daily for food, Virginia tobacco farmers reported that they lost almost $16 daily on food, implying that the total cost of food for H-2 workers in Virginia was $20 to $21 daily or equivalent to the 1980 federal government allowance for travelling employees, who were presumably eating in restaurants.

A Review of Agriculture Statistics Programs of the Bureau of the Census and U.S. Department of Agriculture. (Departments of Commerce and Agriculture, 1983). This report was prepared at the request of OMB to justify the Census of Agriculture. It includes responses from federal agencies that use Census data; these reports indicate how the data is used and what substitute would be used if Census data were unavailable. Most responses stress the virtues of the large Census sample which provides state and county-level data to bench mark annual or biennial surveys based on small samples.

Rosenberg, Howard. "Personal Management of Lettuce Harvest Crews," Iceberg Lettuce Research Program Annual Report 1982-83, pp. 99-133. A 1982 survey of 482 California lettuce workers which found workers whose average age was 33.3; an average 7.2 years of education (two-thirds got all their education in Mexico, where 78 percent were born). Sample workers were employed an average 7.6 months per year and had 11 years experience in lettuce; but did not want to be doing the same job in five years (only 31 percent did) and they would "definitely not" recommend their jobs to their children (52 percent).

Roth, Dennis. Counting Migrant and Seasonal Farmworkers: A Persistent Data Void, Congressional Research Service; 85-797E; July 26, 1985. This report reviews the strengths and weaknesses of the COA, QALS, the CPS-HFWF, the COP, and the MSRTS.

Rural America. Where Have All The Farmworkers Gone?
(Washington, D.C., 1977). This is the report discussed in
Chapter 3 which estimated the number of migrant and seasonal
workers and their dependent sin 1976 on the basis of ES-223
data.

Schlenger, William, Lynn Ordrizek, and Jerome Hallan. "An
Examination of the Methods Used to Estimate the Number of
Migratory Seasonal Farmworkers," The Farmworker Journal,
Winter 1978-79, pp. 21-38. Dated (1971) critique of the major
farmworker data series which concludes that none are totally
reliable and that all tend to underestimate migrants. Notes that
the 1970 HFWF estimate of 196,000 migrants has a 95 percent
confidence interval of 145,000 to 245,000, i.e., the "true"
number of migrants has a 95 percent chance of being between
145,000 to 245,000. In addition to this large standard error, the
report notes that migrant housing units in December may not be
randomly distributed but instead concentrated in Texas, Florida,
California, and Mexico.

Smith, Leslie W. and Robert Coltrane. Hired Farmworkers:
Background and Trends for the Eighties (Washington: USDA,
ERS, RDR Report 32, 1981). Reviews data from Farm Labor
and The Hired Farm Working Force to emphasize: (1) the
growing importance of hired workers as family farm
employment declines (hired were 35 percent of farm
employment in 1980); and (2) the 1970s stability of farm jobs
(average 1.2 million) and farmworkers (2.6 million).

Sosnick, Stephen. Hired Hands: Seasonal Farmworkers in the
United States (Santa Barbara: McNally and Lottin, 1978).

U.S. House of Representatives, Hunger Among Migrant and
Seasonal Farmworkers, , Select Committee on Hunger, Serial
99-22, July 31, 1986. This hearing focuses on migrant
problems in several Eastern states. The opening statement
estimates that there are 800,000 migrant workers and dependents
and 1.9 million seasonals and dependents (Source: HHS,
1980). The report cites 1983 HFWF data for the earnings and
education characteristic of migrant workers, but does not cite
sources for the health-related statistics such as the migrants' 49-
year life expectancy.

U.S. Department of Health and Human Services, Public Health
Service, 1978 Migrant Health Program Target Population
Estimates, April 1980. This report is described as "an
enlightened review of existing, admittedly flawed data," (p. 1).

This report, prepared for Migrant Health by InterAmerica Research Associates, adjusts ES-223 data to estimate the number of migrant farmworkers, migrant dependents, and seasonal farmworkers in 910 counties classified as high impact (more than 4,000 migrants and their dependents) and combined high impact (migrants, their dependents, and seasonals exceed 4,000). The report concludes that there are 800,000 migrants and dependents and 1.87 million seasonal workers in migrant impacted areas, or 2.67 million farmworkers and dependents (N.B. Seasonal workers in non-impact areas and year-round farmworkers are excluded).

USDA, ERS. Domestic Migratory Farmworkers (Washington, USDA AER 121, 1967). This report reviews the characteristics of the estimated 400,000 migrant farmworkers in the mid-1960s. The leitmotiv of the report is that migrants were similar to other farmworkers: Most migrants were young White males who are out of the workforce most of the year. Migrants were interviewed "at their homes" in December, and 40 percent were based in the South, 30 percent in the West, and 30 percent in North and Midwest. Surprisingly, even though migrants left their usual county and home overnight to do farmwork, only 40 percent had more than one farm employer. This report includes a four page discussion of migrant household characteristics: of the estimated 466,000 migrant workers in 1965, about 40 percent were household heads, 24 percent were 14 to 17 year-olds who did migratory farmwork with and without their families, 24 percent were unrelated adults (18 and older) without families, and 12 percent were the wives of migrant household heads. The families of the 180,000 migratory household heads included 721,000 persons or four per household, including 1.7 children under 14. White migrants were concentrated in the North, where 94 percent of all migrants were White and West (86 percent) White; non-Whites were concentrated in the South, where 40 percent of all migrants are non-White.

U.S. House of Representatives, Committee on Education and Labor, Subcommittee on Labor Standards, Hearings on Immigration Reform and Guestworkers, April 11 and May 3, 1984, Washington, D.C.

U.S. House of Representatives, Committee on the Judiciary. Immigration Reform and Control Act of 1986, Conference

Report, 99th Congress, 2nd session, October 14, 1986, Report 99-1000.

U.S. House of Representative. Oversight Hearing on Migrant Education Programs. Committee on Education and Labor, Subcommittee on Agricultural Labor, 1975. This hearing was chaired by William Ford (D-Michigan), the Chief Congressional patron of Migrant Education programs. This hearing explains why Migrant Education developed the MSRTS and began to use MSRTS data to allocate funds in FY 1975.

U.S. Senate. The Migratory Farm Labor Problem in the U.S. Committee on Labor and Public Welfare, Subcommittee on Migratory Labor, Report 1098, 1961. This report summarizes 16 findings and recommendations and then summarizes 11 bills introduced to deal with inter alia, minimum wages, child labor, migrant education, and collective bargaining.

U.S. Senate. Migrant and Seasonal Worker Powerlessness. Subcommittee on Migratory Labor, 1969-70. This 16-volume report was produced by the Senate Subcommittee in 1969-70, when it was chaired by Walter Mondale. The reports include an overview of who migrants are, their legal and union problems, immigration and pesticide issues, and a 3-part hearing entitled "Who is responsible?."

U.S. Senate. Farmworkers in Rural America 1971-72. Subcommittee on Migratory Labor, 1971 and 1972. This 5-part report was produced by the Senate Subcommittee in 1971-72, when it was chaired by Adlai Stevenson. It covers broader issues affecting farmworkers, such as land ownership patterns and the activities of land-grant universities.

U.S. Senate. Violations of Free Speech and the Rights of Labor. Committee on Education and Labor (the LaFollette Committee). This report was issued between 1940 and 1962 in a series of reports which comprised 50 volumes and almost 20,000 pages. The final report was issued in March 1942 as Senate Report 1150.

Whitener, Leslie A. Counting Hired Farmworkers: Some Points to Consider (Washington: USDA, ERS-AER 524, 1984). Similar article published as "A Statistical Portrait of Hired Farmworkers," Monthly Labor Review, June 1984, pp. 49-53. This report contrasts the characteristics of five farmworker occupational codes from the 1980 Census of Population (475, 476, 477, 479, 484) with characteristics of farmworkers in the December 1981 HFWF sample. According to Whitener, the

COP reported that 792,000 farmworkers were employed in March 1980, in these occupational codes, while the HFWF estimated 818,000 employed in March 1981. Farmworkers employed in the March 1981 HFWF are more likely to be year-round White males committed to farmwork than farmworkers who worked sometime in 1981 but not in March. This contrast suggests that the 1980 COP likely missed many migrant and seasonal workers and the farmworker characteristics it reports are most likely traits of year-round farmworkers.

Appendix A

Migrant Farmworker Assistance Programs

COMMUNITY FACILITIES LOAN PROGRAM

U.S. Department of Agriculture

1. Services Provided:

    Loan funds to public bodies, non-profit organizations, and Federally recognized Indian tribes in rural areas of 20,000 or less in population for essential community facility projects for health care, public safety, and public service.

2. Statutory Authority and Regulations:

    (a) Consolidated Farm and Rural Development Act, Section 306; and
    (b) 7 CFR 1942.

3. Contact Person: Anne Williams/Chris Alsop
    Address: Community Facilities Division, Room 6304
                    14th & Independence Avenue, S.W.
                    Washington, D.C. 20250

    Telephone Number: (202) 382-1490
4. Total Federal Funds for FY: 1987, $95.7 million

5. Program Administered by:

    (a) FMHA District offices; and
    (b) State Farmers and Home Administration Offices.

6. Definition of Eligible Migrants:

    Migrant agriculture laborers and their family dependents who establish a temporary residence while performing agriculture work at one or more locations away from the place they call home or home base. (This does not include day-haul agriculture workers whose travels are limited to work areas within one day of their work location.) A home base State is a State which the migrant farm laborer claims as his/her domicile.

7. Program Publications and Directories:

    NA

FARM LABOR HOUSING LOAN AND GRANT PROGRAM

U. S. Department of Agriculture

1. Services Provided:

   The program provides funds for the construction of housing and other ancillary services located at a housing site.

2. Statutory Authority and Regulations:

   (a) The Housing Act of 1949, as amended; and
   (b) 7 CFR 1944.

3. Contact Person: Rebecca W. Johnson
   Address: Multi-Family Housing Division
   Room 5343 South Building
   Washington, D.C. 20250

   Telephone Number: (202) 382-1627

4. Total Federal Funds for FY: 1987, $12 million loans and
   $10 million in grants (431 units
   of rental assistance)

5. Program Administered by:

   (a) Farm Labor Housing Administration state offices; and
   (b) Farm Labor Housing district offices.

6. Definition of Eligible Migrants:

   Migrant agriculture laborer means any agriculture laborer and family dependents who establish a temporary residence while performing agriculture work at one or more locations away from the place he/she calls home or home base. (This does not include day-haul agricultural workers whose travels are limited to work areas within one day of their work location). A home base state which the migrant farm laborer claims as his/her domicile.

7. Program Publications and Directories:

   NA

NATIONAL SCHOOL BREAKFAST PROGRAM (NSBP)

U.S. Department of Agriculture

1. Services Provided:

    The program provides subsidized breakfast service in schools for eligible school children.

2. Statutory Authority and Regulations:

    (a) The Child Nutrition Act of 1966, Section 4, as amended; The National School Lunch Act, Section 9; and
    (b) 7 CFR 220.

3. Contact Person: Diane Berger
    Address: Food and Nutrition Service, USDA
                      3101 Park Center Drive
                      Alexandria, VA 22302

    Telephone Number: (703) 756-3620

4. Total Federal Funds for FY: 1988, $480 million

5. Program Administered by:

    (a) State Education Agencies;
    (b) School Food Authorities (School District); and
    (c) U.S. Department of Agriculture Food and Nutrition Regional Offices

6. Definition of Eligible Migrants:

    Children of migrant families meeting prescribed family size and income eligibility standards can receive free or reduced price benefits if their school elects to participate in the NSBP program. Both public and non-profit private schools may participate.

7. Program Publications and Directories:

    NA

NATIONAL SCHOOL LUNCH PROGRAM (NSLP)

U.S. Department of Agriculture

1. Services Provided:

   The program provides subsidized lunch service in schools for eligible school children.

2. Statutory Authority and Regulations:

   (a) The National School Lunch Act, as amended; and
   (b) 7 CFR 210.

3. Contact Person: Diane Berger
   Address: Food and Nutrition Service, USDA
                  3101 Park Center Drive
                  Alexandria, VA 22302

   Telephone Number: (703) 756-3620

4. Total Federal Funds for FY: 1988, $4 billion and
                               $934 million in commodities.

5. Program Administered by:

   (a) State Education Agencies;
   (b) School Food Authorities (School Districts); and
   (c) USDA Food and Nutrition Regional Offices.

6. Definition of Eligible Migrants:

   Children of migrant families meeting proscribed family size and income eligibility standards can receive free or reduced price benefits if their school elects to participate in the NSLP program. Both public and non-profit private schools may participate.

7. Program Information and Directories:

   "National School Lunch Program," FNS-78, available by request from above address.

SPECIAL MILK PROGRAM (SMP)

U.S. Department of Agriculture

1. Services Provided:

    Provides subsidized or free 1/2 pints of milk for children in participating non-profit private and public schools.

2. Statutory Authority and Regulations:

    (a) The Child Nutrition Act, Section 3, as amended; and
    (b) 7 CFR 215.

3. Contact Person: Diane Berger
    Address: Food and Nutrition Service, USDA
                        3101 Park Center Drive
                        Alexandria, VA 22302

    Telephone Number: (703) 756-3620

4. Total Federal Funds for FY: 1988, $34 million

5. Program Administered by:

    (a) State Education Departments;
    (b) School Food Authorities (School Districts); and
    (c) U.S. Department of Agriculture Food and Nutrition Regional Offices.

6. Definition of Eligible Migrants:

    Children of migrant families meeting prescribed family size and income eligibility standards can receive free or reduced price benefits if their school elects to participate in the SMP program. Both public and non-profit private schools may participate.

7. Program Publications and Directories:

    NA

SPECIAL SUPPLEMENTAL FOOD PROGRAM FOR WOMEN, INFANTS AND
CHILDREN (WIC)
U.S. Department of Agriculture

1. Services Provided:

WIC provides specific nutritious supplemental foods, and nutrition education
to low income pregnant, postpartum and breast-feeding women, infants, and
children to age five, who are determined to be at nutritional risk. Special
program provisions have been implemented in response to the unique needs
of the migrant population. For example, a verification of certification card is
issued to ensure continuity of benefits as migrant participants move from area
to area. In addition, WIC legislation provides for a special migrant set-aside
of not less than nine-tenths of one percent of appropriated funds for services
to eligible migrant populations.

2. Statutory Authority and Regulations:

(a) Child Nutrition Act of 1966, Section 17, as amended; and
(b) CFR 246.

3. Contact Person: Doris M. Dvorscak
   Address: Food and Nutrition Service, USDA
                   3101 Park Center Drive, Room 406
                   Alexandria, VA 22302

   Telephone Number: (703) 756-3730

4. Total Federal Funds for FY: 1987, $1.66 billion

5. Program Administered by:

(a) 87 State agencies, including District of Columbia, Guam, Puerto Rico,
    Virgin Islands and 33 Indian State agencies;
(b) 1675 local agencies overseeing 8,000 service sites; and
(c) Seven FNS Regional offices nationwide.

6. Definition of Eligible Migrants:

Migrant Farmworkers are people whose principal employment is in agriculture
on a seasonal basis, who have been so employed within the last 24 months,
and who establish, for the purposes of such employment, a temporary abode.

7. Program Publications and Directories:

(a) "How WIC Helps," USDA Food and Nutrition Service, Program Aid
    No. 1198 (Spanish only)
(b) "Supplemental Food Programs of the United States Department of
    Agriculture." USDA, Food and Nutrition Service, FNS-235,
    FNS-235-S (English and Spanish)

COLLEGE ASSISTANCE MIGRANT PROGRAM (CAMP)

U.S. Department of Education

1. Services Provided:

    CAMP provides funds to colleges and universities to assist migrant and seasonal farmworkers enrolled as first year undergraduates to make a successful transition from secondary to post-secondary education.

2. Statutory Authority and Regulations:

    (a) Higher Education Act of 1965, Section 418A; and
    (b) 34 CFR 206.

3. Contact Person: Dr. John F. Staehle
    Address: Office of Migrant Education
                         Mail Stop 6275
                         400 Maryland Avenue, S.W.
                         Washington, D.C. 20202

    Telephone Number: (202) 732-4746

4. Federal Funds For: 1987, $1.2 million

5. Program Administered by:

    (a) The Federal Office of Migrant Education; and
    (b) Institutions of higher education; or
    (c) Non-profit organizations in conjunction with colleges or universities.

6. Definition of Eligible Migrants:

    (a) Migrant Farmworker means a seasonal farmworker whose employment required travel that precluded the farmworker from returning to his or her domicile (permanent place of residence) the same day.
    (b) Seasonal Farmworker means a person who within the past 24 months was employed for at least 75 days in farmwork and whose primary employment was in farmwork on a temporary or seasonal basis (that is, not a constant year-round activity).

7. Program Publications or Directories:

    See Appendix A.

HANDICAPPED MIGRATORY AGRICULTURAL AND SEASONAL FARMWORKER VOCATIONAL AND REHABILITATION SERVICES PROGRAM

U. S. Department of Education

1. Services Provided:

    Rehabilitation services to meet the needs of handicapped migrant and seasonal farmworkers and their immediate family members.

2. Statutory Authority and Regulations:

    (a) The Education of the Handicapped Act, Part B, as amended; and
    (b) 34 CFR 375.

3. Contact Person: Frank Caraciolo
    Address: Rehabilitation Services Administration
                      U.S. Department of Education
                      Switzer Building, Room 3320
                      330 C Street, S.W.
                      Washington, D.C. 20202

    Telephone Number: (202) 732-1340

4. Total Federal Funds for FY: 1987, $1,057

5. Program Administered by:

    (a) State Rehabilitation Agencies; and
    (b) Local organizations that have agreements with State Rehabilitation Agencies.

6. Definition of Eligible Migrants:

    Migrant agricultural workers or seasonal farmworkers and members of their immediate family.

7. Program Publications and Directories:

    NA

HIGH SCHOOL EQUIVALENCY PROGRAM (HEP)

U.S. Department of Education

1. Services Provided:

   HEP funds colleges and universities to prepare migrant and seasonal farmworkers to obtain a High School Equivalency Certificate (GED) and to assist with their transition to a higher education institution, job training program for full-time employment.

2. Statutory Authority and Regulations:

   (a) The Higher Education Act, Section 418A; and
   (b) 34 CFR 206.

3. Contact Person: Dr. John F. Staehle
   Address: Office of Migrant Education
                     Mail Stop 6275
                     400 Maryland Avenue, S.W.
                     Washington, D.C. 20202

   Telephone Number: (202) 732-4746

4. Total Federal Funds for FY: 1988, $6.3 million

5. Program Administered by:

   (a) The Federal Office of Migrant Education; and
   (b) Institutions of higher education; or
   (c) Non-profit organizations in conjunction with colleges and universities.

6. Definition of Eligible Migrants:

   (a) Migrant farmworker means a seasonal farmworker whose employment required travel that precluded the farmworker from returning to his or her domicile (permanent place of residence) with in the same day.
   (b) Seasonal farmworker means a person who within the past 24 months was employed for at least 75 days in farmwork and whose primary employment was in farmwork on a temporary or seasonal basis (that is, not a constant year round-activity).

7. Program Publications and Directories:

   See Appendix B.

INTER/INTRASTATE COORDINATION (SECTION 143)

U.S. Department of Education

1. Services Provided:

   (a) Inter/Intrastate Contracts: are intended to improve the coordination of education and support services for migrant children nationwide.
   (b) Migrant Student Record Transfer System (MSRTS): is a contract that provides for a computerized system of transferring academic and health data on migrant children as they move between state and local education agencies.

2. Statutory Authority and Regulations:

   (a) The Education Consolidation and Improvement Act of 1981, Section 143; and
   (b) no regulations.

3. Contact Person: Dr. John F. Staehle
   Address: Office of Migrant Education
                      400 Maryland Avenue, S.W.
                      Washington, D.C. 20202

   Telephone Number: (202) 732-4746

4. Total Federal Funds for FY: 1988, $7.065 million.

5. Program Administered by:

   (a) State Educational Agencies (SEAs); and
   (b) MSRTS Headquarters in Little Rock, AK.

6. Definition of Eligible Migrants:

   In general these contracts serve children of migratory agricultural or fishery workers.

7. Program Publications and Directories:

   For further information write to above address.

STATE BASIC GRANT PROGRAM (SECTION 141)

U.S. Department of Education

1. Services Provided:

    Provides funds to state educational agencies (SEAs) to implement
    instruction and support programs for migrant school-aged children.

2. Statutory Authority and Regulations:\

    (a) The Education Consolidation and Improvement Act of 1981, Section
         141; and
    (b) 34 CFR 201.

3. Contact Person: Dr. John F. Staehle
    Address: Office of Migrant Education, USED
                         Mail Stop 6275
                         400 Maryland Avenue, S.W.
                         Washington, D.C. 20202

    Telephone Number: (202) 732-4746

4. Total Federal Funds for FY: 1988, $257,458.4 million

5. Program Administered by:

    (a) State Educational Agencies (SEAs); and
    (b) Local Education Agencies (School Districts).

6. Definition of eligible migrants:

    In general, the program serves children of currently or formerly (up to five
    years) migratory agricultural fishery workers.

7. Program Publications and Directories:

    See appendix C.

PESTICIDE FARM SAFETY PROGRAM

Environmental Protection Agency

1. Services Provided:

   The staff coordinates the Environmental Protection Agency's activities to protect farmworkers from the hazards of pesticides, develops worker protection standards for farmworkers and applicators for the safe use of pesticides, and develops and distributes pesticide safety materials for farmworkers and applicators.

2. Statutory Authority and Regulations:

   (a) Federal Insecticide, Fungicide, and Rodenticide Act; and
   (b) 40 CFR 70.

3. Contact Person: Patricia Breslin
   Address: Pesticide Farm Safety Staff, TS-757C
                     EPA
                     401 M. Street, S.W.
                     Washington, D.C. 20460

   Telephone Number: (703) 557-7666

4. Total Federal Funds for FY: 1987, $100,000

5. Program Administered by:

   The Federal U.S. Environmental Protection Agency
   (Regional E A offices oversee State enforcement program.)

6. Definition of Eligible Migrants:

   Program serves all agricultural farm laborers.

8. Program Information and Directories:

   (a) Pesticide Safety for Farmworkers, (Bilingual), EPA.
   (b) Pesticide Safety for Non-Certified Mixers, Loaders and Applicators, (Bilingual), EPA.

COMMUNITY SERVICES BLOCK GRANT PROGRAM

U.S. Department of Health and Human Services

1. Services Provided:

    Provides funds for states to provide programs designed to eliminate various causes of poverty.

2. Statutory Authority and Regulations:

    (a) Community Services Block Grant Act of 1981, Section 671, as amended; and
    (b) 45 CFR 16,74 and 96.

3. Contact Person: Jane Checkan
    Address: Office of Community Services
                        Family Support Administration
                        Health and Human Services
                        330 C Street, S.W.
                        Washington, D.C. 20201

    Telephone Number: (202) 475-0391

4. Total Federal Funds for FY: 1987, $335 million

5. Program Administered by:

    State Offices

6. Definition of Eligible Migrants:

    Eligibility is defined separately by State. Previous beneficiaries of funds include low income elderly, children, homeless, and migrants.

7. Program Information and Directories:

    See Appendix D.

MIGRANT HEAD START

U.S. Department of Health and Human Services

1. Services Provided:

   Provides a comprehensive early childhood program for migrant preschool children.

2. Statutory Authority and Regulations:

   (a) Head Start Act, Omnibus Budget Reconciliation Act of 1981, as amended; and
   (b) 45 CFR 1304.

3. Contact Person: Robert Radford
   Address: 400 6th Street, S.W.\
                    Donohoe Building
                    Washington, D.C. 20213

   Telephone Number: (202) 755-7782

4. Total Federal Funds for FY: 1987, $38 million.

5. Program Administered by:

   (a) Federal Migrant Head Start Office; and
   (b) Local non-profit organizations.

6. Definition of Eligible Migrants:

   Migrant pre-school aged children of parents who have changed their place of residence in the last 12 months in search of agricultural work.

7. Program Publications and Directories:

   See Appendix E.

MIGRANT HEALTH PROGRAM
U.S. Department of Health and Human Services

1. Services Provided:

   Provides physician diagnosis and primary health care services for migrant and seasonal farmworkers.

2. Statutory Authority and Regulations:

   (a) Public Health Service Act, Title III, Section 329; and
   (b) 42 CFR 56.

3. Contact Person: Sonia M. Leon Reig
   Address: Office of Migrant Health
                      DPCS/BHCD/DHHS
                      5600 Fishers Lane
                      Room 7A30
                      Rockville, Maryland 20857

   Telephone Number: (301) 443-1153

4. Total Federal Funds for FY: 1987, $45.4 million.

5. Program Administered by:

   (a) Ten Regional Health and Human Service Department Offices; and
   (b) Local health centers.

6. Definition of eligible migrants:

   (a) "Migrant Farmworker" is a person whose principal employment is in agriculture and has been so employed in the last 24 months and has established a temporary abode for such employment.
   (b) "Seasonal Farmworker" is a person whose principal employment is in agriculture and returns to his or her permanent residence after each day's work.

7. Program Publications and Directories:

   "Migrant Health Center Directory" is available by request at the National Migrant Referral Project, Inc., Austin, Texas (512) 447-0770.

   "Health for the Nation's Harvesters", Helen Johnston. A history of the Migrant Health Program at the Federal level.

MIGRANT AND SEASONAL WORKER PROTECTION ACT (MSPA) PROGRAM
U.S. Department of Labor

1. Services Provided:

   (a) Registration: The program insures that migrant and seasonal agricultural workers are properly registered if they are engaging in any farm labor contracting activity and provided the protections under MSPA.
   (b) Enforcement: The Wage and Hour division enforces the provisions of MSPA which include the proper wage payments, safe transportation and healthful housing for all registered farm workers.

2. Statutory Authority and Regulations:

   (a) Migrant and Seasonal Agricultural Worker Protection Act; and
   (b) 29 CFR 500.

3. Contact Person: Gordon Claucherty
   Address: Employment Standards Administration
   Wage and Hour Division
   Division of Farm and Child Labor Programs
   200 Constitution Avenue, NW Rm S-3504
   Washington, D.C. 20210

   Telephone Number: (202) 523-4670

4. Total Federal Funds: NA

5. Program Administered by:

   The Employment Standards Administration, Washington, D.C.

6. Definition of Eligible Migrants:

   (a) Migrant agricultural worker: is an individual who is employed in agricultural employment of a seasonal or other temporary nature, and who is required to be absent overnight from his permanent place of residence.
   (b) Seasonal agricultural worker: is an individual who is employed in agricultural employment of a seasonal or other temporary nature and is not required to be absent overnight from his permanent place of residence.

7. Program Publications and Directories:\

   "Fact Sheet" No. ESA-14; and
   See Appendix F.

MIGRANT AND SEASONAL FARMWORKER PROGRAM (JTPA)

U.S. Department of Labor

1. Services Provided:

   Job training employment opportunities, job search assistance, training related and non-training related supportive services leading to eventual placement in unsubsidized agricultural or non-agricultural employment.

2. Statutory Authority and Regulations:

   (a) Job Training Partnership Act, Section 402; and
   (b) 20 CFR 633.

3. Contact Person: Charles C. Kane
   Address: 200 Constitution Avenue, N.W.
                     Room N 4643
                     Washington, D.C. 20210

   Telephone Number: (202) 535-0500

4. Total Federal Funds for FY: 1987, $57,32 million

5. Program Administered by:

   Public agencies and private non-profit organizations funded through biennial competition on a state of substate basis.

6. Definition of Eligible Migrants:

   Eligibility is limited to migrant and seasonal farmworkers, and their dependents who during any consecutive 12-month period within the 24 month period preceding their application for enrollment received at least 50% of their total earned income or were employed at least 50% of their total work time in farmwork and whose annual family income does not exceed the higher of either the poverty level of 70% of the federal guidelines.

7. Program Publications and Directories:

   See Appendix G.

MIGRANT AND SEASONAL FARMWORKER SERVICES (MSFW)

U.S. Department of Labor

1. Services Provided:

   MSFW provides employment services to domestic U.S. Migrant and Seasonal Farmworkers that include, a complaint system, outreach, job referral, training, and referral to other supportive services.

2. Statutory Authority and Regulations:

   (a) Wagner Peyser Act of 1933, as amended; and
   (b) 20 CFR 651.

3. Contact Person: Gilbert Apodaca
   Address: Employment and Training Administration, USDL
                     200 Constitution Avenue, N.W., Room N-4456
                     Washington, D.C. 20003

   Telephone Number: (202) 535-0163\

4. Total Federal Funds for FY: Not Applicable

5. Program Administered by:

   (a) State Employment Service Agencies; and
   (b) Public agencies and private non-profit organizations funded through biennial competition on a state or substate basis.

6. Definition of Eligible Migrants:

   Migrant and seasonal farmworkers who are unable to return to their residences at the end of their work day.

7. Program Publications and Directories:

   NA

TEMPORARY ALIEN AGRICULTURAL LABOR CERTIFICATION PROGRAM (H2-A)

U.S. Department of Labor

1. Services Provided:

    Provides temporary, foreign farmworkers for employers who meet certification requirements.

2. Statutory Authority and Regulations:

    (a) Immigration, Reform and Control Act of 1986, Section 216; and
    (b) 29 CFR 501.

3. Contact Person: John Hancock
    Address: Division of Foreign Labor Certification
                       200 Constitution Avenue, N.W.
                       Washington, D.C. 20003

    Telephone Number: (202) 535-0166

4. Total Federal Funds for FY: 1987, $3.5 million

5. Program Administered by:

    Division of Foreign Labor Certification, U.S. Department of Labor.

6. Definition of eligible migrants:

    Migrants who work in this program are not U.S. citizens. An employer is eligible for temporary foreign farmworker assistance if: 1) there are no sufficient qualified, able and willing U.S. domestic workers available at the time and place needed, and 2) employment of such foreign farmworkers would not adversely affect similarly employed U.S. domestic workers.

7. Program Publications and Directories:

    NA

FARMWORKERS JUSTICE FUND

Non-Federal Legal Services Organization

1. Services Provided:

   This organization acts as an advocacy group that provides legal and legislative services and provides technical resources for farmworkers.

2. Contact Person: Ron D'Aloisio
   Address: 2001 S. Street, N.W., Suite 210
                     Washington, D.C. 20009

   Telephone Number: (202) 462-8192

3. Program Administered by:

   Headquarter office in Washington, D.C.

4. Definition of Eligible Migrants:

   The   serve all migrant and seasonal farmworkers and their organizations.

5. Program Publication and Directories:

   NA

MIGRANT LEGAL ACTION PROGRAM

Non-Federal Legal Services Organization

1. Services Provided:

    The program operates as a national legal services support center which provides representation to migrant and seasonal farmworkers and provides technical assistance to migrant and legal services field programs nationwide.

2. Contact Person: Roger Rosenthal
    Address: Migrant Legal Action
                        2001 S Street, N.W.
                        Washington, D.C. 20009

    Telephone Number: (202) 462-7744

3. Program Administered by:

    Headquarter office in Washington, D.C.

4. Definition of eligible migrants:

    The program refers eligibility questions to local field programs who determine clients eligibility on the basis of Legal Services Corporation regulations.

5. Program Publications and Directories:

    NA

INTERSTATE MIGRANT EDUCATION COUNCIL/EDUCATION COMMISSION OF THE STATES

Non-Federal Education Organization

1. Services Provided:

   Encourages interstate cooperation and sharing of information in migrant education, conducts policy analysis of pertinent issues and serves as a liaison between State and Federal programs.

2. Contact Person: John P. Perry/Jim L. Gonzalez
   Address: 1860 Lincoln Street
   Suite 300
   Denver, Colorado 80295

   Telephone Number: (303) 830-3680

4. Program Administered by:

   Interstate Migrant Education Council

5. Definition of migrants used to meet eligibility requirements:

   Eligibility is as defined by Office of Migrant Education, USED.

6. Program Publications and Directories:

   NA

NATIONAL GOVERNORS ASSOCIATION

1. Services Provided:

   Assists migrant farmworkers by encouraging States to provide needed
   assistance to migrant farmworkers. Presently the Association is also trying
   to mobilize funds for a directory of protective laws for farmworkers.

2. Contact Person: Fernando L. Alegria, Jr.
   Address: Hall of Stables
                      Suite 250
                      444 North Capitol Street
                      Washington, D.C. 20001

   Telephone Number: (202) 624-5427

3. Program Administered by:

   Washington, D.C. headquarters

4. Definition of eligible migrants:

   NA

5. Program Information and Directories:

   NA

Appendix B

Farm Labor Data

There is both too much and too little farm labor data. The Census of Agriculture and the various migrant assistance programs produce reams of numbers, but many have only an indirect relationship to a clearly-defined migrant population. This appendix summarizes some of the major farm labor data that is available on a state-by-state basis.

There are several caveats to bear in mind when studying this data. First, U.S. totals are those for the states listed; several data sources also count or assist farmworkers in Puerto Rico and other territories, so the U.S. total reported here may not reflect the total number published in the source document. Second, concepts differ between the data sources. The tables have footnotes which explain these conceptual differences, and Chapter 2 elaborates further on them. Third, there are no decimal points in the tables. Although the data can be presented with decimal points, readers of this book will appreciate why we could not apply a false sense of precision to farm labor market estimates.

Table A 1.1

Farm Sales and Labor Expenditures Reported in the 1982 Census of Agriculture

State	Ag Products Sold($000)	Per Dist	Crop & Live Lab Expend -($000)-	Per Dist	Crop & Live Cont Expend -($000)-	Per Dist	Crop & Live Tot Expend -($000)-	Per Dist	Crop Products Sold($000)	Per Dist	Crop-Per of Ag Prod Solc	Crop Lab Expend -($000)-	Per Dist
Alabama	1,704,160	1%	87,505	1%	6,515	1%	94,020	1%	606,916	1%	36%	36,363	1%
Alaska	11,399	0%	1,925	0%	91		2,016		6,010		53%		
Arizona	1,526,915	1%	156,401	2%	31,239	3%	187,640	2%	806,847	1%	53%	140,090	3%
Arkansas	2,826,497	2%	164,885	2%	7,807	1%	172,692	2%	1,356,618	2%	48%	121,334	2%
California	12,491,442	9%	1,819,323	22%	413,766	37%	2,233,089	23%	8,158,494	13%	65%	1,681,436	31%
Colorado	2,940,897	2%	129,180	2%	11,205	1%	140,385	1%	847,323	1%	29%	48,556	1%
Connecticut	285,324	0%	44,278	1%	1,462		45,740		102,058		36%	10,056	
Delaware	370,562	0%	17,893	0%	1,296		19,189		110,276		30%	8,084	
Florida	3,522,103	3%	480,444	6%	201,298	18%	681,742	7%	2,518,959	4%	72%	472,408	9%
Georgia	2,767,679	2%	149,524	2%	14,181	1%	163,705	2%	1,180,988	2%	43%	90,000	2%
Hawaii	558,608	0%	146,277	2%	3,455		149,732	2%	456,069	1%	82%	128,196	2%
Idaho	2,231,605	2%	147,584	2%	18,733	2%	166,317	2%	1,160,742	2%	52%	100,754	2%
Illinois	7,313,529	6%	225,820	3%	8,024	1%	233,844	2%	5,092,452	8%	70%	134,405	2%
Indiana	4,226,930	3%	152,061	2%	8,469	1%	160,530	2%	2,439,409	4%	58%	69,711	1%
Iowa	9,828,932	7%	222,146	3%	8,358	1%	230,504	2%	4,143,086	7%	42%	91,027	2%
Kansas	6,190,861	5%	153,404	2%	8,123	1%	161,527	2%	2,143,047	3%	35%	72,429	1%
Kentucky	2,376,882	2%	166,518	2%	11,809	1%	178,327	2%	1,358,309	2%	57%	97,667	2%
Lousiana	1,406,458	1%	107,345	1%	4,255		111,600	1%	994,976	2%	71%	82,883	2%
Maine	399,412	0%	44,906	1%	3,436		48,342	1%	142,834	1%	36%	25,068	
Maryland	1,029,244	1%	68,545	1%	4,473		73,018	1%	339,430	1%	33%	23,583	
Massachuse	281,436	0%	43,109	1%	2,604		45,713	1%	139,428	1%	50%	16,708	
Michigan	2,588,317	2%	186,312	2%	14,445	1%	200,757	2%	1,364,665	2%	53%	99,052	2%
Minnesota	5,939,629	5%	207,615	2%	9,127	1%	216,742	2%	2,671,482	4%	45%	88,420	2%
Mississippi	1,918,486	1%	137,400	2%	4,714		142,114	1%	1,122,471	2%	59%	100,190	2%
Missouri	3,606,856	3%	141,232	2%	9,235	1%	150,467	2%	1,546,664	2%	43%	64,475	1%
Montana	1,547,160	1%	84,462	1%	6,016	1%	90,478	1%	759,171	1%	49%	40,777	1%
Nebraska	6,625,742	5%	167,515	2%	7,757	1%	175,272	2%	2,379,811	4%	36%	69,122	1%
Nevada	202,581	0%	20,438		1,124		21,562		72,582		36%	7,468	

Farm Sales and Labor Expenditures Reported in the 1982 Census of Agriculture

State	Ag Products Sold($000)	Per Dist	Crop & Live Lab Expend -($000)-	Per Dist	Crop & Live Cont Expend -($000)-	Per Dist	Crop & Live Tot Expend -($000)-	Per Dist	Crop Products Sold($000)	Per Dist	Crop-Per Ag Prod Sold	Crop Lab Expend -($000)-	Per Dist
New Hampshire	102,520	0%	13,626	0%	578		14,204		26,207		26%	4,252	
New Jersey	435,966	0%	62,380	1%	11,632	1%	74,012	1%	322,038	1%	74%	45,104	1%
New Mexico	850,562	1%	61,178	1%	8,930	1%	70,108	1%	232,230		27%	29,509	1%
New York	2,426,936	2%	246,022	3%	12,778	1%	258,800	3%	657,719	1%	27%	90,809	2%
North Carolina	3,500,750	3%	245,364	3%	20,747	2%	266,111	3%	1,898,109	3%	54%	183,623	3%
North Dakota	2,294,326	2%	76,864	1%	3,961		80,825	1%	1,759,871	3%	77%	63,230	1%
Ohio	3,387,461	3%	166,229	2%	9,585	1%	175,814	2%	1,863,940	3%	55%	64,907	1%
Oklahoma	2,530,061	2%	98,335	1%	11,864	1%	110,199	1%	827,694	1%	33%	36,320	1%
Oregon	1,640,590	1%	179,512	2%	14,696	1%	194,208	2%	935,456	2%	57%	108,212	2%
Pennsylvania	2,848,207	2%	224,174	3%	12,249	1%	236,423	2%	751,988	1%	26%	44,645	1%
Rhode Island	30,376	0%	5,559	0%	118		5,677		18,139		60%	992	
South Carolina	968,554	1%	79,734	1%	9,581	1%	89,315	1%	601,018	1%	62%	53,336	1%
South Dakota	2,478,111	2%	67,414	1%	3,333	1%	70,747	1%	842,980	1%	34%	21,804	
Tennessee	1,683,852	1%	109,251	1%	8,187	1%	117,438	1%	848,819	1%	50%	54,912	1%
Texas	8,936,363	7%	480,462	6%	88,334	8%	568,796	6%	3,025,698	5%	34%	274,717	5%
Utah	555,428	0%	42,066	0%	3,399		45,465		130,233		23%	8,306	
Vermont	369,402	0%	28,865	0%	435		29,300		20,054		5%	2,280	
Virginia	1,606,915	1%	126,893	2%	10,340	1%	137,233	1%	629,303	1%	39%	58,303	1%
Washington	2,831,159	2%	313,100	4%	33,501	3%	346,601	4%	1,714,741	3%	61%	262,910	5%
West Virginia	242,127	0%	20,340	0%	2,371		22,711		57,203		24%	9,946	
Wisconsin	4,854,582	4%	279,154	3%	9,988	1%	289,142	3%	943,422	2%	19%	61,758	1%
Wyoming	606,327	0%	40,613	0%	4,148		44,761		128,106		21%	8,645	
United States	131,900,223	100%	8,441,180	100%	1,103,773	100%	9,544,953	100%	62,256,087	100%	47%	5,408,782	100%

Source: Bureau of the Census, Census of Agriculture, 1982; a farm is a place which sold farm products worth $1000 or more in 1982
Labor expenditures include gross wages or salaries paid to family and nonfamily hired workers and supervisors, bonuses,
social security and other payroll taxes, and expenditures for fringe benefits
Contract labor expenditures include wages paid to hired workers and supervisors, payroll taxes and fringe benefits, and contractor or crew le
Seasonal workers were employed less than 150 days on the responding farm; a worker employed on two farms is counted twice in this data

Table A 1.2

Hired Worker Employment Reported in the 1982 Census of Agriculture

State	Crop and Livestock Workers	Per Dist	Crop & Live Seasonal Workers	Seasonal Per of Tot Workers	Per Dist	Crop & Live Regular Workers	Regular Per of Tot Workers	Per Dist	Total Crop Workers	Per Dist
Alabama	74,523	2%	51,368	69%	1%	23,250	31%	2%	36,759	1%
Alaska			1,116	100%						
Arizona	62,904	1%	33,671	53%	1%	30,000	47%	3%	54,469	2%
Arkansas	86,765	2%	48,087	55%	1%	38,800	45%	4%	56,302	2%
California	979,874	20%	640,750	65%	16%	339,510	35%	36%	921,620	28%
Colorado	62,076	1%	35,031	56%	1%	27,256	44%	3%	41,758	1%
Connecticut	19,647		9,633	49%		10,052	51%	1%	15,515	
Delaware	10,191		6,103	60%		4,124	40%		6,162	
Florida	235,381	5%	129,316	55%	3%	106,106	45%	11%	209,967	6%
Georgia	109,998	2%	73,940	67%	2%	36,088	33%	4%	77,756	2%
Hawaii	17,406		7,190	41%		10,233	59%	1%	15,635	1%
Idaho	94,347	2%	61,913	65%	2%	32,650	35%	3%	70,063	2%
Illinois	164,547	3%	116,349	71%	3%	48,340	29%	5%	125,147	4%
Indiana	110,792	2%	82,813	72%	2%	31,961	28%	3%	70,856	2%
Iowa	206,130	4%	146,776	73%	4%	55,627	27%	6%	115,455	4%
Kansas	85,730	2%	54,137	63%	1%	31,640	37%	3%	56,383	2%
Kentucky	287,064	6%	247,808	86%	6%	39,352	14%	4%	219,532	7%
Lousiana	59,316	1%	32,940	55%	1%	26,504	45%	3%	42,010	1%
Maine	39,882	1%	30,784	77%	1%	9,120	23%	1%	30,875	1%
Maryland	46,710	1%	29,082	62%	1%	17,708	38%	2%	32,931	1%
Massachusetts	22,400		12,305	55%		10,206	45%	1%	16,710	1%
Michigan	150,151	3%	109,807	73%	3%	40,408	27%	4%	117,093	4%
Minnesota	182,779	4%	126,435	69%	3%	56,417	31%	6%	101,243	3%
Mississippi	79,979	2%	43,268	54%	1%	36,790	46%	4%	53,643	2%
Missouri	118,632	2%	84,369	71%	2%	34,394	29%	4%	53,904	2%
Montana	45,344	1%	23,715	52%	1%	21,810	48%	2%	27,686	1%
Nebraska	96,464	2%	60,743	63%	2%	35,827	37%	4%	60,546	2%
Nevada	8,812		3,886	44%		4,964	56%	1%	4,666	2%

Hired Worker Employment Reported in the 1982 Census of Agriculture

State	Crop and Livestock Workers	Per Dist	Crop & Live Seasonal Workers	Seasonal Per of Tot Workers	Per Dist	Crop & Live Regular Workers	Regular Per of Tot Workers	Per Dist	Total Crop Workers	Per Dist
New Hampshire	7,904		4,544	55%		3,663	45%		4,896	
New Jersey	36,072	1%	20,065	56%	1%	16,038	44%	2%	32,307	1%
New Mexico	34,977	1%	21,716	61%	1%	13,958	39%	1%	19,458	1%
New York	135,023	3%	77,351	57%	2%	57,880	43%	6%	87,091	3%
North Carolina	271,293	6%	233,420	86%	6%	38,109	14%	4%	237,769	7%
North Dakota	59,628	1%	30,015	50%	1%	29,768	50%	3%	50,508	2%
Ohio	117,720	2%	85,841	73%	2%	31,957	27%	3%	73,521	2%
Oklahoma	80,886	2%	52,715	65%	1%	28,291	35%	3%	42,085	1%
Oregon	165,254	3%	120,408	73%	3%	44,993	27%	5%	140,616	4%
Pennsylvania	91,740	2%	62,582	68%	2%	29,414	32%	3%	46,759	1%
Rhode Island	11,813		934	8%		10,898	92%	1%	11,444	
South Carolina	68,631	1%	52,863	74%	1%	18,883	26%	2%	56,326	2%
South Dakota	54,590	1%	32,277	59%	1%	22,542	41%	2%	28,328	1%
Tennessee	208,727	4%	139,244	66%	4%	70,148	34%	7%	149,811	5%
Texas	236,500	5%	173,712	73%	4%	62,976	27%	7%	119,069	4%
Utah	34,241	1%	23,903	70%	1%	10,433	30%	1%	15,492	1%
Vermont	29,376	1%	8,429	29%		21,082	71%	2%	19,772	1%
Virginia	132,913	3%	89,534	67%	2%	43,435	33%	5%	86,876	3%
Washington	259,202	5%	230,121	89%	6%	29,195	11%	3%	228,461	7%
West Virginia	67,191	1%	18,359	27%		48,876	73%	5%	53,159	2%
Wisconsin	166,471	3%	115,026	69%	3%	51,506	31%	5%	49,015	1%
Wyoming	20,069		9,982	49%		10,194	51%	1%	9,458	
United States	4,839,112	100%	3,906,376	81%	100%	943,943	19%	100%	3,287,263	100%

Source: Bureau of the Census, Census of Agriculture, 1982; a farm is a place which sold farm products worth $1000 or mor
Labor expenditures include gross wages or salaries paid to family and nonfamily hired workers and supervisors, bonuses,
social security and other payroll taxes, and expenditures for fringe benefits
Contract labor expenditures include wages paid to hired workers and supervisors, payroll taxes and fringe benefits, and contract
Seasonal workers were employed less than 150 days on the responding farm; a worker employed on two farms is counted twic

Table A 1.3

Hired Worker Employment Reported in the 1982 Census of Agriculture

State	Per of Crop and Live Workers	Crop Seasonal Workers	Per Dist	Per of Tot Crop Workers	Crop Reg Per Dist	Live Seasonal Workers	Per Dist	Live Reg Per Dist
Alabama	49%	19,036	1%	52%	3%	5,527	1%	3%
Alaska								
Arizona	87%	28,139	1%	52%	5%	3,670	1%	
Arkansas	65%	23,307	1%	41%	6%	5,805	2%	2%
California	94%	607,601	22%	66%	56%	25,491	7%	3%
Colorado	67%	21,837	1%	52%	4%	7,335	2%	1%
Connecticut	79%	7,211		46%	1%	1,748		
Delaware	60%	3,136		51%	1%	1,098		
Florida	89%	112,293	4%	53%	18%	8,432	2%	1%
Georgia	71%	49,795	2%	64%	5%	8,127	2%	2%
Hawaii	90%	6,537		42%	2%	1,135		
Idaho	74%	44,532	2%	64%	5%	7,119	2%	1%
Illinois	76%	85,081	3%	68%	7%	8,274	2%	3%
Indiana	64%	46,661	2%	66%	4%	7,766	2%	3%
Iowa	56%	77,510	3%	67%	7%	17,682	5%	6%
Kansas	66%	32,510	1%	58%	4%	7,767	2%	2%
Kentucky	76%	189,009	7%	86%	5%	8,829	2%	5%
Louisiana	71%	18,809	1%	45%	4%	3,303	1%	1%
Maine	77%	24,221	1%	78%	1%	2,466	1%	1%
Maryland	71%	20,143	1%	61%	2%	4,920	1%	1%
Massachusetts	75%	8,746		52%	1%	2,242	1%	
Michigan	78%	85,743	3%	73%	6%	9,058	2%	2%
Minnesota	55%	62,983	2%	62%	7%	18,157	5%	5%
Mississippi	67%	21,842	1%	41%	6%	4,989	1%	2%
Missouri	45%	28,731	1%	53%	5%	9,221	2%	5%
Montana	61%	12,027		43%	3%	6,151	2%	1%
Nebraska	63%	35,480	1%	59%	4%	10,761	3%	2%
Nevada	53%	1,469		31%	1%	1,767		2%

Hired Worker Employment Reported in the 1982 Census of Agriculture

State	Per of Crop and Live Workers	Crop Seasonal Workers	Per Dist	Per of Tot Crop Workers	Crop Reg Per Dist	Live Seasonal Workers	Per Dist	Live Reg Per Dist
New Hampshire	62%	2,304		47%		1,071		
New Jersey	90%	17,762	1%	55%	3%	1,493		
New Mexico	56%	9,456	2%	49%	2%	3,956	1%	1%
New York	65%	48,511	8%	56%	7%	19,300	5%	2%
North Carolina	88%	207,800	8%	87%	5%	8,140	2%	2%
North Dakota	85%	23,272	1%	46%	5%	2,532	1%	1%
Ohio	62%	51,333	2%	70%	4%	9,769	3%	3%
Oklahoma	52%	20,963	1%	50%	4%	7,169	2%	3%
Oregon	85%	101,025	4%	72%	7%	5,402	1%	2%
Pennsylvania	51%	32,328	1%	69%	3%	14,983	4%	3%
Rhode Island	97%	673		6%	2%	127		
South Carolina	82%	40,324	1%	72%	3%	2,881	1%	1%
South Dakota	52%	12,332		44%	3%	6,546	2%	2%
Tennessee	72%	85,870	3%	57%	11%	6,207	2%	5%
Texas	50%	84,227	3%	71%	6%	28,134	7%	8%
Utah	45%	9,710		63%	1%	4,651	1%	1%
Vermont	67%	2,693		14%	3%	4,003	1%	
Virginia	65%	52,529	2%	60%	6%	9,088	2%	3%
Washington	88%	205,624	8%	90%	4%	6,358	2%	2%
West Virginia	79%	5,707		11%	9%	1,424		1%
Wisconsin	29%	37,069	1%	76%	2%	39,560	10%	7%
Wyoming	47%	3,487		37%	1%	4,223	1%	1%
United States	68%	2,729,388	100%	83%	100%	386,068	100%	100%

Source: Bureau of the Census, Census of Agriculture, 1982; a farm is a place which sold farm products wor
Labor expenditures include gross wages or salaries paid to family and nonfamily hired workers and superviso
social security and other payroll taxes, and expenditures for fringe benefits
Contract labor expenditures include wages paid to hired workers and supervisors, payroll taxes and fringe ben
Seasonal workers were employed less than 150 days on the responding farm; a worker employed on two farm

Table A 1.4

Vegetable(016) Employment and Wages Reported in the 1982 Census of Agriculture

State	Vegetable Farm Employer	Per Dist	Veg Labor Expend -($000)-	Per Dist	Large Vegetable Employers	Per Dist	Regular Vegetabl Dist Workers	Per Dist	Reg-Per of Tot Workers	Seasonal Vegetable Workers	Per of Tot Workers	Per Dist	Veg Cont Labor Expend -($000)-	Per Dist
Alabama	345	2%	1,368	6%	3		169		8%	2,058	92%	1%	869	
Alaska														
Arizona	128	1%	38,648	6%	45	2%	2,683	4%	18%	11,854	82%	4%	18,022	8%
Arkansas	202	1%	750		1		119		11%	947	89%		197	
California	1,845	13%	345,703	52%	817	42%	33,985	49%	29%	84,275	71%	32%	96,195	43%
Colorado	209	1%	6,545	1%	23	1%	703	1%	20%	2,894	80%	1%	1,814	1%
Connecticut	106	1%	743		4		117		13%	762	87%		78	
Delaware	36		2,787		8		197		27%	545	73%		151	
Florida	895	6%	92,200	14%	225	12%	12,659	18%	21%	48,263	79%	18%	40,518	18%
Georgia	377	3%	2,676		14	1%	261	1%	10%	2,416	90%	1%	2,406	1%
Hawaii	165	1%	3,200		21	1%	432	1%	56%	337	44%		449	
Idaho	81	1%	1,960		8		265		10%	2,290	90%	1%	650	
Illinois	164	1%	5,716	1%	15	1%	437	1%	14%	2,704	86%	1%	857	
Indiana	68		257				48		8%	523	92%		61	
Iowa	256	2%	2,808		18	1%	261	1%	10%	2,315	90%	1%	2,366	1%
Kansas	50		286		1		38		18%	174	82%		335	
Kentucky	113	1%	180		1		29		7%	365	93%		34	
Lousiana	232	2%	448				72		5%	1,249	95%		60	
Maine	89	1%	725		3		100		15%	573	85%		39	
Maryland	203	1%	1,751		11	1%	151	1%	11%	1,266	89%		493	
Massachusetts	297	2%	2,363		10	1%	384	1%	20%	1,522	80%	1%	417	
Michigan	1,001	7%	19,335	3%	89	5%	1,565	2%	9%	15,684	91%	6%	3,641	2%
Minnesota	248	2%	2,937		12	1%	320	1%	12%	2,309	88%	1%	435	
Mississippi	177	1%	246				43		8%	478	92%		17	
Missouri	50		334				55		19%	242	81%		80	
Montana	18		28							70	100%			
Nebraska	24		74							82	100%			
Nevada	6		167		1		9		6%	131	94%			

Vegetable(016) Employment and Wages Reported In the 1982 Census of Agriculture

State	Vegetable Farm Employer	Per Dist	Veg Labor Expend -($000)-	Per Dist	Large Vegetable Employer	Per Dist	Regular Vegetable Workers	Per Dist	Reg-Per of Tot Workers	Seasonal Vegetable Workers	Seasonal Per of Tot Workers	Per Dist	Veg Cont Labor Expend -($000)-	Per Dist
New Hampshire	53		469		4		62		10%	529	90%	2%	3,714	2%
New Jersey	653	5%	16,164	2%	98	2%	2,612	5%	31%	5,832	69%	1%	4,663	2%
New Mexico	182	1%	5,007	1%	29	1%	424	2%	13%	2,794	87%	3%	2,981	1%
New York	976	7%	25,012	4%	104	4%	2,532	5%	22%	9,114	78%	1%	1,719	1%
North Carolina	514	4%	2,954		15		453	1%	11%	3,698	89%	1%		
North Dakota	12													
Ohio	510	4%	10,679	2%	38	2%	1,017	2%	13%	6,660	87%	3%	3,535	2%
Oklahoma	60		415		2		83		16%	433	84%		414	
Oregon	464	3%	11,264	2%	71	2%	885	4%	8%	9,778	92%	4%	1,792	1%
Pennsylvania	343	2%	3,775	1%	15	1%	559	1%	19%	2,367	81%	1%	1,677	1%
Rhode Island	24		184		1		28		24%	87	76%		12	
South Carolina	275	2%	4,412	1%	20	1%	589	1%	14%	3,518	86%	1%	2,778	1%
South Dakota	22													
Tennessee	339	2%	2,016		8		198		9%	2,021	91%	1%	658	1%
Texas	889	6%	24,979	4%	80	4%	2,792	4%	19%	11,974	81%	5%	20,324	9%
Utah	123	1%	835		5		170		18%	762	82%		306	
Vermont	71		201				59		20%	243	80%		2	
Virginia	248	2%	3,175		8	1%	422	1%	18%	1,982	82%	1%	1,422	1%
Washington	589	4%	13,721	2%	63	2%	1,139	3%	8%	13,150	92%	5%	5,273	2%
West Virginia	15		22				7		13%	49	88%		18	
Wisconsin	501	4%	8,366	1%	39	1%	901	2%	20%	3,689	80%	1%	2,433	1%
Wyoming														
United States	14,248	100%	667,885	100%	1,930	100%	70,034	100%	21%	265,008	79%	100%	223,905	100%

Source: Bureau of the Census, Census of Agriculture, 1982; a farm is a place which sold farm products worth $1000 or more in 1982

Labor expenditures include gross wages or salaries paid to nonfamily hired, bonuses, payroll taxes, and the cost of fringe benefits

Seasonal workers were employed less than 150 days on the responding farm; a worker employed on two farms is counted twice in this data

Regular workers were employed on the responding farm more than 150 days;

Large farm employers reported paying $50,000 or more in 1982; large contract labor farms paid $20,000 or more.

Table A 1.5
Fruit and Nut (017) Employment and Wages Reported in the 1982 Census of Agriculture

State	Fruit and Nut Farm Ers	Per Dist	Fruit & Nut Lab Expend -($000)-	Per Dist	Large Fruit and Nut Employers	Per Dist	Fruit & Ni Reg-Per Regular Workers	Per of Tot Workers	Per Dist	Fruit & Ni Seasonal Seasonal Workers	Per of Tot Workers	Per Dist	Fruit & Ni Seasonal Contract Lab Expend -($000)-	Per Dist
Alabama	247	1%	967		3		141	9%		1,497	91%		240	
Alaska														
Arizona	255	1%	16,004	1%	40	1%	2,882	36%	3%	5,193	64%	1%	3,168	1%
Arkansas	156		1,017		5		110	7%		1,420	93%		258	
California	19,935	46%	625,043	54%	1,934	49%	59,007	12%	55%	422,461	88%	52%	237,702	57%
Colorado	249	1%	1,908	1%	4		222	10%		2,099	90%		306	
Connecticut	117		2,214		10		224	20%		870	80%		190	
Delaware	28		232		1		79	43%		104	57%			
Florida	3,180	7%	112,088	10%	390	10%	11,010	25%	10%	32,182	75%	4%	110,201	26%
Georgia	675	2%	9,692	1%	48	1%	935	9%	1%	9,182	91%	1%	2,720	1%
Hawaii	395	1%	39,338	3%	21	1%	2,516	36%	2%	4,481	64%	1%	1,525	
Idaho	196		3,931		9		525	9%		5,586	91%	1%	234	
Illinois	222	1%	3,017		19		436	12%		3,330	88%		163	
Indiana	61		242				20	4%		476	96%		94	
Iowa	248	1%	2,913		15		279	11%		2,323	89%			
Kansas	21		204				14	6%		232	94%		111	
Kentucky	70		338		1		66	14%		398	86%		3	
Lousiana	170		757		2		38	3%		1,131	97%		55	
Maine	382	1%	6,269	1%	26	1%	504	7%		6,718	93%	1%	2,422	1%
Maryland	132		4,066		14		424	25%		1,306	75%		890	
Massachuset	351	1%	10,326	1%	52	1%	943	18%	1%	4,246	82%	1%	1,416	1%
Michigan	2,190	5%	32,567	3%	164	4%	2,635	7%	2%	32,973	93%	4%	4,981	1%
Minnesota	131		1,973		7		166	8%		1,869	92%		83	
Mississippi	131		295				42	7%		565	93%		89	
Missouri	145		2,672		17		276	12%		2,106	88%		299	
Montana	76		233				10	1%		792	99%		238	
Nebraska	9		73							37	100%			
Nevada	23		137				20	69%		9	31%			

Fruit and Nut (017) Employment and Wages Reported in the 1982 Census of Agriculture

State	Fruit and Nut Farm Ers	Fruit & Nut Lab Expend ($000)-	Per Dist	Large Fruit and Nut Employers	Per Dist	Fruit & Nt Regular Workers	Fruit & Nt Reg-Per of Tot Workers	Per Dist	Fruit & Nt Seasonal Workers	Per of Tot Workers	Per Dist	Contract Lab Expend ($000)-	Per Dist
New Hampshire	109	2,955		16		386	24%		1,200	76%		311	
New Jersey	287	12,768	1%	58	1%	1,268	16%	1%	6,471	84%	1%	5,811	1%
New Mexico	323	5,455	1%	8		595	28%	1%	1,541	72%		124	
New York	1,713	35,382	4%	181	5%	2,849	10%	3%	25,516	90%	3%	4,916	1%
North Carolina	403	2,900	1%	8		324	9%		3,307	91%		952	
North Dakota	2												
Ohio	502	4,636	1%	22	1%	467	9%	1%	4,984	91%	1%	567	
Oklahoma	151	703		4		136	15%		785	85%		113	
Oregon	1,837	30,708	4%	143	4%	1,892	3%	2%	53,865	97%	7%	4,698	1%
Pennsylvania	840	17,404	2%	77	2%	1,815	15%	2%	10,262	85%	1%	3,651	1%
Rhode Island	30	244		1		43	27%		118	73%		39	
South Carolina	2												
South Dakota													
Tennessee	91	273											
Texas	956	11,392	2%	33	1%	1,508	27%	1%	4,021	73%	1%	4,100	1%
Utah	240	1,683	1%	7		210	5%		3,669	95%		319	
Vermont	81	1,373		9		122	10%		1,126	90%		75	
Virginia	391	10,803	1%	45	1%	1,296	20%	1%	5,113	80%	1%	2,912	1%
Washington	4,748	116,785	11%	495	12%	8,956	6%	8%	147,081	94%	18%	20,759	5%
West Virginia	120	6,651	1%	29	1%	684	21%	1%	2,561	79%	1%	1,845	
Wisconsin	390	9,458	1%	61	2%	663	12%	1%	4,733	88%	1%	858	
Wyoming													
United States	43,011	1,150,089	100%	3,979	100%	106,738	12%	100%	819,939	88%	100%	419,438	100%

Source: Bureau of the Census, Census of Agriculture, 1982; a farm is a place which sold farm products worth $1000 or more in 1982
Labor expenditures include gross wages or salaries paid to nonfamily hired, bonuses, payroll taxes, and the cost of fringe benefits
Seasonal workers were employed less than 150 days on the responding farm; a worker employed on two farms is counted twice in this data
Regular workers were employed on the responding farm more than 150 days;
Large farm employers reported paying $50,000 or more in 1982; large contract labor farms paid $20,000 or more.

Table A 1.6

Horticulture Specialty(018) Employment and Wages Reported in the 1982 Census of Agriculture

State	Hort Specialty Farm Ers	Per Dist	Hort Specialty Expend -($000)-	Per Dist	Large Hort Specialty Employers	Per Dist	Regular Workers	Reg-Per of Tot Workers	Per Dist	Seasonal Workers	Seasonal Per of Tot Workers	Per Dist	Contract Labor Expend -($000)-	Per Dist
Alabama	246	2%	13,196	1%	43	1%	1,683	46%	2%	1,941	54%	1%	965	3%
Alaska														
Arizona	77	1%	5,840	1%	20	1%	403	32%	1%	860	68%	1%	651	2%
Arkansas	111	1%	2,424		11		303	27%	1%	803	73%	1%	119	
California	1,933	13%	269,571	28%	709	22%	23,918	46%	24%	27,877	54%	20%	7,475	21%
Colorado	277	2%	19,902	2%	81	3%	1,889	41%	2%	2,708	59%	2%	536	2%
Connecticut	252	2%	20,013	2%	48	1%	2,275	50%	2%	2,299	50%	2%	758	2%
Delaware	55		1,215		7		171	44%		217	56%		94	
Florida	1,805	12%	124,189	13%	418	13%	15,480	48%	15%	16,450	52%	12%	6,953	20%
Georgia	247	2%	7,429	1%	42	1%	761	28%	1%	1,962	72%	1%	546	2%
Hawaii	284	2%	8,517	1%	45	1%	1,047	57%	1%	794	43%	1%	433	1%
Idaho	134	1%	2,169		9		294	19%		1,258	81%	1%	70	1%
Illinois	330	2%	28,798	3%	97	3%	2,846	50%	3%	2,793	50%	2%	869	2%
Indiana	163	1%	7,685	1%	27	1%	799	41%	1%	1,171	59%	1%	107	1%
Iowa	238	2%	12,298	1%	42	1%	1,306	49%	1%	1,353	51%	1%	62	1%
Kansas	123	1%	4,879	1%	19	1%	572	51%	1%	551	49%		34	
Kentucky	203	1%	4,346		22	1%	643	31%	1%	1,437	69%	1%	77	
Lousiana	202	1%	4,854		25	1%	641	45%	1%	770	55%	1%	222	1%
Maine	83	1%	1,575		7		202	30%		469	70%		25	
Maryland	244	2%	10,341	1%	46	1%	1,289	40%	1%	1,902	60%	1%	474	1%
Massachuse	242	2%	9,985	1%	41	1%	1,291	45%	1%	1,552	55%	1%	340	1%
Michigan	644	4%	32,384	3%	125	4%	3,214	32%	3%	6,712	68%	5%	836	2%
Minnesota	248	2%	12,131	1%	49	2%	1,151	39%	1%	1,832	61%	1%	231	1%
Mississippi	121	1%	2,794		11		470	42%		653	58%		23	
Missouri	241	2%	9,477	1%	38	1%	951	41%	1%	1,391	59%	1%	174	1%
Montana	54		1,207		4		101	14%		624	86%		9	
Nebraska	72		1,972		8		232	28%		595	72%		15	
Nevada	10		178		2		25	14%		153	86%			

Horticulture Specialty(018) Employment and Wages Reported in the 1982 Census of Agriculture

State	Hort Specialty Farm Ers	Per Dist	Hort Specialty Expend ($000)	Per Dist	Large Hort Specialty Employers	Per Dist	Regular Workers	Reg-Per of Tot Workers	Per Dist	Seasonal Workers	Seasonal Per of Tot Workers	Per Dist	Contract Labor Expend ($000)	Per Dist
New Hampshire	60		2,618		12		256	47%		293	53%		33	
New Jersey	459	3%	17,929	2%	85	3%	2,011	46%	2%	2,395	54%	2%	1,178	3%
New Mexico	57		2,974		10		360	41%		528	59%		248	1%
New York	719	5%	26,773	3%	127	4%	2,664	38%	3%	4,310	62%	3%	683	2%
North Carol	438	3%	11,863	1%	52	2%	1,642	36%	2%	2,887	64%	2%	676	2%
North Dako	40		854		4		98	21%		358	79%		16	
Ohio	659	4%	37,648	4%	159	5%	3,973	43%	4%	5,220	57%	4%	261	1%
Oklahoma	148	1%	12,394	1%	28	1%	967	38%	1%	1,585	62%	1%	505	1%
Oregon	601	4%	38,727	4%	118	4%	3,409	27%	3%	9,119	73%	7%	1,187	3%
Pennsylvan	902	6%	90,234	9%	248	8%	9,544	51%	9%	9,014	49%	7%	3,258	9%
Rhode Islan	43		3,930		14		363	54%		307	46%		25	
South Caro	84	1%	4,119		24	1%	440	44%		564	56%		50	
South Dako	24		732		3		81	27%		221	73%		63	
Tennessee	345	2%	14,781	2%	51	2%	1,744	46%	2%	2,084	54%	2%	2,064	6%
Texas	637	4%	35,300	4%	105	3%	3,810	45%	4%	4,652	55%	3%	2,080	6%
Utah	81	1%	2,873		12		472	33%		957	67%	1%	66	
Vermont	56		551		3		108	28%		272	72%		9	
Virginia	302	2%	10,885	1%	54	2%	1,367	42%	1%	1,919	58%	1%	328	1%
Washington	371	2%	22,871	2%	78	2%	2,271	24%	2%	7,247	76%	5%	356	1%
West Virgin	84	1%	1,666		9		197	42%		271	58%		6	
Wisconsin	366	2%	11,654	1%	43	1%	1,424	36%	1%	2,521	64%	2%	142	
Wyoming	14		171		1		25	18%		117	82%		10	
United State	15,129	100%	970,916	100%	3,236	100%	101,183	42%	100%	137,968	58%	100%	35,342	100%

Source: Bureau of the Census, Census of Agriculture, 1982; a farm is a place which sold farm products worth $1000 or more in 1982

Labor expenditures include gross wages or salaries paid to nonfamily hired; bonuses, payroll taxes, and the cost of fringe benefits

Seasonal workers were employed less than 150 days on the responding farm; a worker employed on two farms is counted twice in this dat

Regular workers were employed on the responding farm more than 150 days;

Large farm employers reported paying $50,000 or more in 1982; large contract labor farms paid $20,000 or more.

Table A 1.7

Fruit, Vegetable, and Hort Specialty Employment and Wages Reported in the 1982 Census of Agriculture

State	Total FVH Farm Ers	FVH Lab Expend ($000)	Per Dist	FVH Large Employer	Per Dist	FVH Regular Workers	Per Dist	FVH Reg-Per of Tot Worker	Per Dist	FVH Seasonal Workers	Seasonal Per of To Workers	Per Dist	Contract Lab Expend ($000)	Per Dist
Alabama	838	15,531	1%	49	1%	1,993		27%		5,496	73%		2,074	
Alaska														
Arizona	460	60,492	1%	105	2%	5,968	1%	25%		17,907	75%	2%	21,841	3%
Arkansas	469	4,191		17		532		14%		3,170	86%		574	
California	23,713	1,240,317	33%	3,460	44%	116,910	38%	18%		534,613	82%	42%	341,372	50%
Colorado	735	28,355	1%	108	1%	2,814	1%	27%		7,701	73%	1%	2,656	
Connecticut	475	22,970	1%	62	1%	2,616	1%	40%		3,931	60%		1,026	
Delaware	119	4,234		16		447		34%		866	66%		245	
Florida	5,880	328,477	8%	1,033	12%	39,149	11%	29%		96,895	71%	14%	157,672	23%
Georgia	1,299	19,797	2%	104	1%	1,957		13%		13,560	87%	1%	5,672	1%
Hawaii	844	51,055	1%	87	1%	3,995	1%			5,612			2,407	
Idaho	411	8,060	1%	26		1,084		11%		9,134	89%	1%	954	
Illinois	716	37,531	1%	131	2%	3,719	1%	30%		8,827	70%	1%	1,889	
Indiana	292	8,184		27		867		29%		2,170	71%		262	
Iowa	742	18,019	1%	75	1%	1,846	1%	24%		5,991	76%		2,428	
Kansas	194	5,369		20		624		39%		957	61%		480	
Kentucky	386	4,864		24		738		25%		2,200	75%		114	
Lousiana	604	6,059		27		751		19%		3,150	81%		337	
Maine	554	8,569		36		806		9%		7,760	91%	1%	2,486	
Maryland	579	16,158	1%	71	1%	1,864	1%	29%		4,474	71%		1,857	
Massachusetts	890	22,674	1%	103	1%	2,618	1%	26%		7,320	74%	1%	2,173	1%
Michigan	3,835	84,286	5%	378	3%	7,414	4%	12%		55,369	88%	3%	9,458	1%
Minnesota	627	17,041	1%	68	1%	1,637	1%	21%		6,010	79%	1%	749	
Mississippi	429	3,335		11		555		25%		1,696	75%		129	
Missouri	436	12,483	1%	55	1%	1,282		26%		3,739	74%		553	
Montana	148	1,468		4		111		7%		1,486	93%		247	
Nebraska	105	2,119		8		232		25%		714	75%		15	
Nevada	39	482		3		54		16%		293	84%			

Fruit, Vegetable, and Hort Specialty Employment and Wages Reported in the 1982 Census of Agriculture

State	Total FVH Farm Ers	Per Dist	FVH Lab Expend ($000)	Per Dist	FVH Large	Per Dist Employer	FVH Regular Workers	Per Dist	Reg-Per of Tot Worker	FVH Seasonal Workers	Per Dist	Seasonal Per of To Workers	Contract Lab Expend ($000)	Per Dist
New Hampshire	222		6,042		32		704		26%	2,022		74%	344	
New Jersey	1,399	2%	46,861	2%	241	3%	5,891	2%	29%	14,698	1%	71%	10,703	2%
New Mexico	562	1%	13,436		47	1%	1,379		22%	4,863		78%	5,035	1%
New York	3,408	5%	87,167	3%	412	5%	8,045	3%	17%	38,940	3%	83%	8,580	1%
North Carolina	1,355	2%	17,717	1%	75	1%	2,419	1%	20%	9,892	1%	80%	3,347	
North Dakota	54		854		4		98		21%	358		79%	16	
Ohio	1,671	2%	52,963	2%	219	2%	5,457	2%	24%	16,864	1%	76%	4,363	1%
Oklahoma	359		13,512		34		1,186		30%	2,803		70%	1,032	
Oregon	2,902	4%	80,699	3%	332	4%	6,186	2%	8%	72,762	6%	92%	7,677	1%
Pennsylvania	2,085	3%	111,413	4%	340	4%	11,918	4%	36%	21,643	2%	64%	8,586	1%
Rhode Island	97		4,358		16		434		46%	512		54%	76	
South Carolina	359		8,531		44		1,029		20%	4,082		80%	2,828	
South Dakota	48		732		3		81		27%	221		73%	63	
Tennessee	775	1%	17,070	1%	59	1%	1,942	1%	32%	4,105		68%	2,722	
Texas	2,482	3%	71,671	3%	218	2%	8,110	3%	28%	20,647	2%	72%	26,504	4%
Utah	444	1%	5,391		24		852		14%	5,388		86%	691	
Vermont	208		2,125		12		289		15%	1,641		85%	86	
Virginia	941	1%	24,863	1%	107	1%	3,085	1%	25%	9,014	1%	75%	4,662	1%
Washington	5,708	8%	153,377	5%	636	7%	12,366	4%	7%	167,478	14%	93%	26,388	4%
West Virginia	219		8,339		38		888		24%	2,881		76%	1,869	
Wisconsin	1,257	2%	29,478	1%	143	2%	2,988	1%	21%	10,943	1%	79%	3,433	1%
Wyoming	14		171		1		25		18%	117		82%	10	
United States	72,388	100%	2,788,890	100%	9,145	100%	277,955	100%	19%	1,222,915	100%	81%	678,685	100%

Source: Bureau of the Census, Census of Agriculture, 1982; a farm is a place which sold farm products worth $1000 or more in 1982

Labor expenditures include gross wages or salaries paid to nonfamily hired , bonuses, payroll taxes, and the cost of fringe benefits

Seasonal workers were employed less than 150 days on the responding farm; a worker employed on two farms is counted twice in this data

Regular workers were employed on the responding farm more than 150 days;

Large farm employers reported paying $50,000 or more in 1982; large contract labor farms paid $20,000 or more.

Table A 1.8
Cash Grains (011) Employment and Wages Reported in the 1982 Census of Agriculture

State	Cash Grain Farm Ers	Per Dist	Cash Grain Expend ($000)	Per Dist	Large Cash Grain Employers	Per Dist	Regular Workers	Reg-Per of Tot Workers	Per Dist	Seasonal Workers	Seasonal Per of Tot Workers	Per Dist	Contract Labor Expend ($000)	Per Dist
Alabama	2,408	1%	10,878	1%	20	1%	1,564	21%	1%	5,779	79%	1%	519	1%
Alaska	78		0										0	
Arizona			667		2		80	30%		190	70%		30	
Arkansas	6,572	3%	103,447	8%	506	15%	11,692	43%	8%	15,731	57%	3%	2,282	3%
California	1,666	1%	46,141	4%	248	7%	3,738	23%	3%	12,402	77%	2%	4,183	6%
Colorado	2,616	1%	19,725	2%	49	1%	2,184	24%	2%	7,048	76%	2%	2,315	3%
Connecticut	11		46										0	
Delaware	488		2,715		8		18	24%		58	76%		86	1%
Florida	516		3,775		9		377	19%	2%	1,580	81%	1%	356	1%
Georgia	2,914	1%	17,594	1%	59	2%	502	22%		1,776	78%	1%	851	1%
Hawaii	8		1,154		7		81	45%		100	55%		0	
Idaho	2,931	1%	23,788	2%	72	2%	2,526	19%	2%	10,865	81%	2%	2,594	4%
Illinois	23,911	11%	117,412	9%	219	7%	11,962	14%	8%	74,371	86%	12%	3,895	6%
Indiana	13,037	6%	61,183	5%	131	4%	6,298	15%	4%	36,149	85%	6%	2,384	3%
Iowa	21,329	10%	77,062	6%	102	3%	8,518	11%	6%	69,172	89%	11%	4,105	6%
Kansas	14,063	7%	61,246	5%	104	3%	6,775	19%	5%	29,297	81%	5%	4,158	6%
Kentucky	4,820	2%	23,450	2%	52	2%	2,794	12%	2%	21,193	88%	3%	869	1%
Lousiana	3,473	2%	36,719	3%	115	3%	4,298	36%	3%	7,641	64%	1%	842	1%
Maine	20		61							70	100%		0	
Maryland	1,197	1%	8,831	1%	25	1%	1,070	24%	1%	3,473	76%	1%	326	1%
Massachuse	3		7							17	100%		0	
Michigan	6,248	3%	21,436	2%	42	1%	2,322	10%	2%	20,246	90%	3%	1,856	3%
Minnesota	14,361	7%	61,510	5%	105	3%	6,784	13%	5%	47,486	87%	8%	2,953	4%
Mississippi	3,730	2%	43,287	3%	175	5%	5,907	40%	4%	9,006	60%	1%	960	1%
Missouri	9,082	4%	51,715	4%	124	4%	5,967	24%	4%	19,145	76%	3%	2,723	4%
Montana	3,744	2%	31,058	2%	64	2%	3,768	34%	3%	7,424	66%	1%	2,119	3%
Nebraska	11,648	5%	58,095	5%	102	3%	6,133	16%	4%	31,831	84%	5%	3,385	5%
Nevada	25		283		1		24	38%		40	63%		2	

Cash Grains (011) Employment and Wages Reported in the 1982 Census of Agriculture

State	Cash Grain Farm Ers	Per Dist	Cash Grain Expend ($000)	Per Dist	Large Cash Grain Employers	Per Dist	Regular Workers	Reg-Per of Tot Workers	Per Dist	Seasonal Workers	Seasonal Per of Tot Workers	Per Dist	Contract Labor Expend ($000)	Per Dist
New Hamps			0										0	
New Jersey	400		2,162		5		285	18%		1,280	82%		103	
New Mexico	469		2,632		3		344	26%		1,005	74%		324	
New York	1,001	1%	7,061	1%	26	1%	825	23%	1%	2,796	77%	2%	266	1%
North Carol	3,931	2%	18,922	1%	48	1%	2,304	14%	2%	14,537	86%	2%	797	3%
North Dako	9,820	5%	48,337	4%	70	2%	6,232	25%	4%	18,497	75%	3%	2,402	3%
Ohio	9,839	5%	35,745	3%	76	2%	4,036	15%	3%	22,418	85%	4%	2,061	4%
Oklahoma	5,511	3%	20,390	2%	14		2,678	19%	2%	11,508	81%	2%	2,518	1%
Oregon	1,400	1%	17,619	1%	65	2%	1,582	22%	1%	5,592	78%	1%	926	1%
Pennsylvani	1,656	1%	6,736	1%	21	1%	878	16%	1%	4,555	84%	1%	431	1%
Rhode Island	1		0										0	
South Carol	2,701	1%	14,023	1%	32	1%	2,021	20%	1%	8,084	80%	1%	491	1%
South Dakota	4,449	2%	18,483	1%	23	1%	2,427	19%	2%	10,628	81%	2%	1,383	2%
Tennessee	4,167	2%	20,896	2%	48	1%	2,724	19%	2%	11,907	81%	2%	792	1%
Texas	10,279	5%	95,489	8%	261	8%	10,680	29%	8%	25,920	71%	4%	10,223	15%
Utah	525		1,645		2		165	9%		1,624	91%		188	
Vermont	3		14										0	
Virginia	1,859	1%	10,072	1%	23	1%	1,192	19%	1%	5,115	81%	1%	851	1%
Washington	2,964	1%	46,343	4%	206	6%	3,788	23%	3%	12,917	77%	2%	1,616	2%
West Virgin	154		561		2		87	19%		360	81%		26	
Wisconsin	2,742	1%	17,728	1%	54	2%	1,889	14%	1%	11,561	86%	2%	753	1%
Wyoming	378		2,061		6		255	22%		892	78%		369	1%
United State	215,159	100%	1,270,305	100%	3,326	100%	142,036	19%	100%	610,802	81%	100%	70,321	100%

Source: Bureau of the Census, Census of Agriculture, 1982; a farm is a place which sold farm products worth $1000 or more in 1982

Labor expenditures include gross wages or salaries paid to nonfamily hired, bonuses, payroll taxes, and the cost of fringe benefits

Seasonal workers were employed less than 150 days on the responding farm; a worker employed on two farms is counted twice in this da

Regular workers were employed on the responding farm more than 150 days;

Large farm employers reported paying $50,000 or more in 1982; large contract labor farms paid $20,000 or more.

Table A 1.9

Other Field (013) Employment and Wages Reported in the 1982 Census of Agriculture

State	Other Field (013)				Large Field (013)				Reg-Per	Seasonal			Contract Labor	
	Farm Ers	Per Dist	Expend ($000)	Per Dist	Employers	Per Dist	Regular Workers	Per Dist	Workers of Tot Workers	Seasonal Workers	Per Dist	Per of Tot Workers	Expend ($000)	Per Dist
Alabama	1,760	1%	15,603	1%	49	2%	2,028	2%	25%	6,019		75%	1,477	1%
Alaska														
Arizona	1,065	1%	46,097	4%	260	8%	4,519	4%	36%	8,076		64%	6,288	4%
Arkansas	917	1%	11,744	1%	63	2%	1,234	1%	27%	3,389		73%	415	
California	2,821	2%	149,233	14%	481	15%	13,119	12%	28%	33,908		72%	30,357	22%
Colorado	936	1%	8,952	1%	41	1%	864	1%	15%	4,937		85%	1,064	1%
Connecticut	185		5,854	1%	12		619	1%	17%	3,096		83%	9	
Delaware	32		1,306		7		91		15%	526		85%	162	
Florida	941	1%	62,702	6%	75	2%	4,485	4%	28%	11,443		72%	31,473	22%
Georgia	3,206	3%	29,508	3%	90	3%	3,324	3%	14%	20,144		86%	2,404	2%
Hawaii	138		81,272	8%	19	1%	5,022		86%	825		14%	697	
Idaho	2,029	2%	44,931	4%	216	7%	4,211	5%	18%	18,575		82%	8,489	6%
Illinois	294		920		2		79		12%	597		88%	34	
Indiana	1,350	1%	3,760		9		393		6%	5,880		94%	107	
Iowa	295		375				59		5%	1,026		95%	34	
Kansas	481		3,039		10		384		23%	1,288		77%	351	
Kentucky	28,794	24%	57,240	5%	35	1%	6,126	6%	4%	152,749		96%	6,119	4%
Louisiana	2,026	2%	41,191	4%	200	6%	4,732	4%	38%	7,599		62%	1,192	1%
Maine	1,078	1%	14,721	1%	64	2%	1,246	1%	7%	16,039		93%	566	
Maryland	1,628	1%	5,048		5		782	1%	7%	11,045		93%	345	
Massachusetts	239		1,391		9		164		12%	1,200		88%	71	
Michigan	1,382	1%	8,426	1%	45	1%	899	1%	13%	5,937		87%	1,034	1%
Minnesota	1,369	1%	13,072	1%	62	2%	1,180	1%	14%	7,076		86%	1,231	1%
Mississippi	2,150	2%	50,410	5%	294	9%	6,594	6%	39%	10,421		61%	1,756	1%
Missouri	1,074	1%	4,344		7		538	1%	11%	4,193		89%	712	1%
Montana	695	1%	4,191		10	1%	639		22%	2,249		78%	819	
Nebraska	359		4,338		20	1%	514		27%	1,408		73%	401	
Nevada	318		5,311		19	1%	539	1%	35%	1,012		65%	378	1%

Other Field (013) Employment and Wages Reported in the 1982 Census of Agriculture

State	Other Field (013) Farm Ers	Per Dist	Field (013) Expend ($000)-	Per Dist	Large Field (013) Employers	Per Dist	Regular Workers	Per Dist	Reg-Per of Tot Workers	Seasonal Workers	Per Dist	Seasonal Per of Tot Workers	Contract Labor Per Dist	Expend ($000)-	Per Dist
New Hamps	172		315		5		46		25%	141		75%		42	
New Jersey	261	1%	1,627				206		17%	1,029		83%		300	
New Mexico	925	1%	7,606	1%	21	1%	1,079	1%	27%	2,962		73%	1%	1,031	1%
New York	1,401	1%	9,360	1%	56	2%	1,077	1%	18%	4,896	1%	82%	1%	872	1%
North Carol	17,419	14%	122,604	11%	259	8%	13,036	12%	7%	172,012	23%	93%	8%	11,220	8%
North Dakot	473		10,690	1%	48	1%	863	1%	18%	4,064	1%	82%	1%	804	1%
Ohio	2,178	2%	3,895		4		307		3%	8,680	1%	97%		221	
Oklahoma	1,043	1%	4,840		8		515		11%	4,010	1%	89%	1%	1,182	1%
Oregon	1,470	1%	22,360	2%	107	3%	1,876	2%	15%	10,771	1%	85%	2%	2,163	2%
Pennsylvani	1,233	1%	4,067		10		522		11%	4,150	1%	89%	1%	164	
Rhode Island	42		399		2		62		30%	147		70%		9	
South Carol	2,604	2%	24,701	2%	66	2%	3,087	3%	11%	25,965	3%	89%	3%	1,904	1%
South Dako	257		1,018		4		156		17%	740		83%		58	
Tennessee	14,098	12%	21,664	2%	25	1%	1,847	2%	3%	63,876	9%	97%	2%	1,731	1%
Texas	8,244	7%	70,342	7%	226	7%	8,150	8%	21%	31,517	4%	79%	8%	14,391	10%
Utah	674	1%	2,258		7		300		13%	1,967		87%		296	
Vermont	218		481		1		69		7%	965		93%		23	
Virginia	6,878	6%	22,657	2%	25	1%	2,889	3%	7%	36,321	5%	93%	3%	2,147	2%
Washington	1,280	1%	44,108	4%	173	5%	3,471	3%	16%	18,897	3%	84%	3%	2,832	2%
West Virgin	623	1%	535				46		2%	1,795		98%		54	
Wisconsin	1,947	2%	17,075	2%	80	2%	1,522	1%	11%	11,869	2%	89%	1%	630	2%
Wyoming	423		3,663		11		443		20%	1,789		80%		745	1%
United State	121,498	100%	1,071,444	100%	3,242	100%	105,976	100%	12%	749,769	100%	88%	100%	140,821	100%

Source: Bureau of the Census, Census of Agriculture, 1982; a farm is a place which sold farm products worth $1000 or more in 1982
Labor expenditures include gross wages or salaries paid to nonfamily hired, bonuses, payroll taxes, and the cost of fringe benefits
Seasonal workers were employed less than 150 days on the responding farm; a worker employed on two farms is counted twice in this da[ta]
Regular workers were employed on the responding farm more than 150 days;
Large farm employers reported paying $50,000 or more in 1982; large contract labor farms paid $20,000 or more.

198

Table A 1.10

General Crop (019) Employment and Wages Reported in the 1982 Census of Agriculture

State	Gen Crop (019) Farm Ers	Per Dist	Gen Crop (019) Expend ($000)	Per Dist	Large Gen Crop (019) Employers	Per Dist	Regular Workers	Reg-Per of Tot Workers	Per Dist	Seasonal Workers	Seasonal Per of Tot Workers	Per Dist	Contract Labor Expend ($000)	Per Dist
Alabama	506	2%	3,206	1%	12	1%	513	23%	2%	1,742	77%	1%	228	
Alaska														
Arizona	138	1%	9,213	3%	30	3%	763	28%	2%	1,966	72%	1%	1,282	2%
Arkansas	223	1%	1,026		5		137	12%		1,017	88%	1%	68	1%
California	907	4%	115,019	38%	364	33%	10,497	28%	33%	26,678	72%	18%	23,676	44%
Colorado	441	2%	4,049	1%	15	1%	431	17%	1%	2,151	83%	1%	1,261	2%
Connecticut	41		96				25	17%		126	83%		20	
Delaware	26		544		4		42	20%		164	80%			
Florida	346	2%	4,919	2%	8	1%	485	18%	2%	2,179	82%	1%	6,805	13%
Georgia	1,538	7%	19,756	7%	81	7%	2,397	22%	8%	8,647	78%	6%	1,805	3%
Hawaii	11		83											
Idaho	821	4%	11,498	4%	46	4%	1,385	19%	4%	5,958	81%	4%	2,600	5%
Illinois	288	1%	1,375		5		136	10%	1%	1,286	90%	1%	50	
Indiana	464	2%	1,347		5		167	6%	1%	2,462	94%	2%	142	
Iowa	389	2%	1,189		4		198	13%	1%	1,321	87%	1%	71	
Kansas	453	2%	2,469	1%	8	1%	270	22%	1%	968	78%	1%	177	
Kentucky	2,378	10%	8,756	3%	13	1%	1,189	8%	4%	12,867	92%	9%	579	1%
Louisiana	160	1%	1,357		8	1%	168	29%	1%	419	71%		15	
Maine	114		216				36	9%		352	91%		17	
Maryland	229	1%	1,176		2		218	16%	1%	1,151	84%	1%	137	
Massachusetts	70		336		2		63	23%		209	77%			
Michigan	564	2%	4,563	2%	19	2%	511	11%	2%	4,191	89%	3%	252	
Minnesota	697	3%	3,551	1%	5	1%	451	16%	1%	2,411	84%	2%	395	1%
Mississippi	205	1%	2,994	1%	15	1%	350	33%	1%	719	67%	1%	102	
Missouri	447	2%	1,282		2		189	10%	1%	1,654	90%	1%	102	
Montana	287	1%	1,896	1%	4		236	21%	1%	868	79%	1%	182	
Nebraska	456	2%	2,486	1%	5		261	15%	1%	1,527	85%	1%	247	
Nevada	19		911		4		98	44%		124	56%		277	1%

General Crop (019) Employment and Wages Reported in the 1982 Census of Agriculture

State	Gen Crop (019) Farm Ers	Per Dist	Gen Crop (019) Expend ($000)	Per Dist	Large Gen Crop (019) Employers	Per Dist	Regular Workers	Reg-Per of Tot Workers	Per Dist	Seasonal Workers	Seasonal Per of Tot Workers	Per Dist	Contract Labor Expend ($000)	Per Dist
New Hampshire	40		111							141	100%		4	
New Jersey	90		962		7	1%	144	16%		755	84%	1%	230	1%
New Mexico	153	1%	1,830	1%	10	1%	221	26%	1%	626	74%	1%	579	1%
New York	406	2%	3,767	1%	22	2%	462	20%	1%	1,879	80%	1%	382	1%
North Carol	1,376	6%	18,651	6%	87	8%	2,272	17%	7%	11,359	83%	8%	2,176	4%
North Dakot	161	1%	895	1%	2		155	31%		353	69%		82	
Ohio	719	3%	2,796	1%	5		319	9%	1%	3,371	91%	2%	352	1%
Oklahoma	950	4%	4,400	1%	9	1%	521	16%	2%	2,642	84%	2%	812	2%
Oregon	450	2%	13,845	5%	88	8%	1,176	9%	4%	11,900	91%	8%	1,532	3%
Pennsylvani	616	3%	3,130	1%	12	1%	470	19%	1%	1,980	81%	1%	104	
Rhode Island	11		62				20	59%		14	41%		4	
South Carol	273	1%	4,662	2%	23	2%	539	20%	2%	2,193	80%	1%	291	1%
South Dako	266	1%	723	1%	6		92	11%		743	89%	1%	73	1%
Tennessee	1,167	5%	4,299	1%	6	1%	495	8%	2%	5,982	92%	4%	468	1%
Texas	1,649	7%	17,109	6%	71	6%	1,859	23%	6%	6,143	77%	4%	4,183	8%
Utah	214	1%	694	1%	2		75	9%		731	91%		4	
Vermont	36		90				16	16%		87	84%		9	
Virginia	615	3%	3,212	1%	5	1%	451	18%	1%	2,079	82%	1%	670	1%
Washington	462	2%	10,012	3%	57	5%	745	11%	2%	6,332	89%	4%	1,027	2%
West Virgin	221	1%	188				22	3%		671	97%		31	
Wisconsin	623	3%	4,011	1%	19	2%	450	14%	1%	2,696	86%	2%	261	2%
Wyoming	141	1%	1,569	1%	7	1%	151	18%		689	82%		227	
United State	22,869	100%	302,352	100%	1,098	100%	31,908	18%	100%	146,560	82%	100%	54,132	100%

Source: Bureau of the Census, Census of Agriculture, 1982; a farm is a place which sold farm products worth $1000 or more in 1982

Labor expenditures include gross wages or salaries paid to nonfamily hired, bonuses, payroll taxes, and the cost of fringe benefits

Seasonal workers were employed less than 150 days on the responding farm; a worker employed on two farms is counted twice in this da

Regular workers were employed on the responding farm more than 150 days;

Large farm employers reported paying $50,000 or more in 1982; large contract labor farms paid $20,000 or more.

Table A 1.11
Average Annual Wages Reported in Farm Labor: 1980, 1985-87

Region	Average annual fieldworker wages				Percent change			Average annual livestock wages				Per Change
	1980	1985	1986	1987	1980-87	1986-87	1985-86	1980	1985	1986	1987	80-87
Northeast I	$2.96	$4.02	$4.50	$4.64	57%	12%	3%	$2.75	$3.47	$3.72	$3.84	40%
Northeast 2	$3.31	$3.99	$4.38	$5.19	57%	10%	18%	$3.25	$3.69	$3.77	$3.91	20%
Appalachia	$3.01	$3.75	$4.00	$4.06	35%	7%	2%	$3.25	$3.95	$4.06	$4.46	37%
Appalachia	$3.05	$3.76	$3.62	$3.74	23%	-4%	3%	$3.11	$3.93	$4.23	$4.08	31%
Southeast	$2.86	$3.29	$3.51	$3.69	29%	7%	5%	$3.06	$3.82	$4.18	$4.32	41%
Florida	$4.03	$4.51	$4.69	$4.92	22%	4%	5%	$3.79	$4.46	$4.57	$4.80	27%
Lake	$3.37	$3.88	$4.37	$4.59	36%	13%	5%	$2.82	$3.51	$3.45	$3.74	33%
Cornbelt 1	$3.48	$4.17	$4.43	$4.69	35%	6%	6%	$3.40	$4.37	$4.32	$4.42	30%
Cornbelt 2	$3.45	$4.29	$4.16	$4.20	22%	-3%	1%	$3.39	$4.00	$4.03	$4.34	28%
Delta	$3.18	$3.77	$3.78	$3.84	21%	0%	2%	$3.30	$4.19	$4.36	$4.16	26%
North Plain	$3.43	$4.54	$4.63	$4.53	32%	2%	-2%	$3.40	$4.39	$4.57	$4.11	21%
South Plain	$3.17	$3.98	$4.51	$4.42	39%	13%	-2%	$3.24	$4.16	$4.47	$4.54	40%
Mountain 1	$3.26	$3.78	$4.47	$4.10	26%	18%	-8%	$3.12	$3.46	$3.78	$3.61	16%
Mountain 2	$3.45	$3.89	$4.46	$4.44	29%	15%	0%	$3.17	$4.47	$5.69	$3.95	25%
Mountain 3	$3.22	$4.28	$4.49	$4.32	34%	5%	-4%	$2.89	$4.24	$4.30	$4.25	47%
Pacific	$3.91	$4.64	$4.44	$5.24	34%	-4%	18%	$3.84	$4.90	$5.25	$5.37	40%
California	$4.26	$5.12	$5.11	$5.35	26%	0%	5%	$4.28	$5.13	$5.50	$5.81	36%
Hawaii	$4.80	$6.51	$6.46	$7.63	59%	-1%	18%	$4.88				
United State	$3.55	$4.30	$4.51	$4.69	32%	5%	4%	$3.25	$4.04	$4.26	$4.29	32%

Source: USDA, Farm Labor

Fieldworkers are nonsupervisory workers engaged in the the preparation, planting, caring, and harvesting of crops and harvesting of crops. Hourly wages are calculated by grouping all the fieldworkers who are paid hourly wages (say 100), reporting the total hours worked during the survey week by all of these workers (6 per day for 5 days is 3000), and listing the gross wages paid to these workers ($12,000). This calculation procedure yields an average fieldworker wage for each reporting farm; note that the $4 average hourly wage on one farm in this example may reflect the presence of equal groups of $3 and $5 per hour workers. This calculated wage procedure weights most the wages paid to workers employed the most hours during the survey week.

Northeast I: CT,ME, MA, NH, NY, RI, VT; Northeast 2: DE, MD, NJ,PA; Appalachian 1: NC, VA; Appalachian 2: KY, TN, WV; Southeast: AL, GA, SC; Lake: MI, MN, WI; Cornbelt: IL, IN, OH; Cornbelt 2: IA, MO; Delta: AR, LA, MS; Northern Plains: KS, NE, ND, SD; Southern Plains: OK, TX; Mountain 1: ID, MT, WY; Mountain 2: CO, NV, UT; Mountain 3: AZ, NM; Pacific: OR, WA.

Table A1.12

Average Annual Wages Reported by Farm Labor for Field and Livestock workers: 1980-1987

Region	1980	Per of US Avg	1985 US Av.	Per of US Av.	1986 US Ave	Per of US Ave	1987	Per of US Ave	Percentage Change 1980-87	1985-86	1986-87	Ranking Region	1980-87
Northeast I	$2.75	85%	$3.78	90%	$4.17	94%	$3.78	90%	37%	10%	-9%	Mountain 3	47%
Northeast 2	$3.25	100%	$3.98	94%	$4.17	94%	$3.90	93%	20%	5%	-6%	Northeast I	37%
Appalachia 1	$3.25	100%	$3.81	90%	$4.02	91%	$3.81	90%	17%	6%	-5%	Hawaii	33%
Appalachia 2	$3.11	96%	$3.82	91%	$3.79	86%	$3.82	91%	23%	-1%	1%	North Plains	32%
Southeast	$3.06	94%	$3.46	82%	$3.73	84%	$3.46	82%	13%	8%	-7%	Lake	31%
Florida	$3.79	117%	$4.50	107%	$4.66	105%	$4.50	107%	19%	4%	-3%	Mountain 2	30%
Lake	$2.82	87%	$3.69	87%	$3.91	88%	$3.69	88%	31%	6%	-6%	South Plains	26%
Cornbelt 1	$3.40	105%	$4.24	100%	$4.38	99%	$4.24	101%	25%	3%	-3%	Cornbelt 1	25%
Cornbelt 2	$3.39	104%	$4.16	99%	$4.10	93%	$4.16	99%	23%	-1%	1%	Appalachia 2	23%
Delta	$3.30	102%	$3.88	92%	$4.05	92%	$3.88	92%	18%	4%	-4%	Cornbelt 2	23%
North Plains	$3.40	105%	$4.50	107%	$4.61	104%	$4.50	107%	32%	2%	-2%	Pacific	22%
South Plains	$3.24	100%	$4.07	96%	$4.49	102%	$4.07	97%	26%	10%	-9%	Northeast 2	20%
Mountain 1	$3.12	96%	$3.60	85%	$4.15	94%	$3.60	86%	15%	15%	-13%	California	20%
Mountain 2	$3.17	98%	$4.12	98%	$5.19	117%	$4.12	98%	30%	26%	-21%	Florida	19%
Mountain 3	$2.89	89%	$4.26	101%	$4.43	100%	$4.26	101%	47%	4%	-4%	Delta	18%
Pacific	$3.84	118%	$4.69	111%	$4.52	102%	$4.69	111%	22%	-4%	4%	Appalachia 1	17%
California	$4.28	132%	$5.12	121%	$5.17	117%	$5.12	122%	20%	1%	-1%	Mountain 1	15%
Hawaii	$4.88	150%	$6.48	154%	$6.43	145%	$6.48	154%	33%	-1%	1%	Southeast	13%
United States	$3.25	100%	$4.22	100%	$4.42	100%	$4.21	100%	30%	5%	-5%	United States	30%

Source: USDA, Farm Labor

Fieldworkers are nonsupervisory workers engaged in the the preparation, planting, caring, and harvesting of crops; livestock workers are general laborers on livestock farms. Hourly wages are calculated by grouping all of the workers who are paid hourly wages (say 100), reporting the total hours worked during the survey week by all of these workers (6 per day for 5 days is 3000), and listing the gross wages paid to these workers ($12,000). This calculation procedure yields an average hourly wage for each reporting farm; note that the $4 average hourly wage on the farm in this example may reflect the presence of equal groups of $3 and $5 per hour workers. This calculated wage procedure weights most the wages paid to workers employed the most hours during the survey week.

Northeast I: CT,ME, MA, NH, NY, RI, VT; Northeast 2: DE, MD, NJ,PA; Appalachian 1: NC, VA; Appalachian 2: KY, TN, WV; Southeast: AL, GA, SC; Lake: MI, MN, WI; Cornbelt: IL, IN, OH; Cornbelt 2: IA, MO; Delta: AR, LA, MS; Northern Plains: KS, NE, ND, SD; Southern Plains: OK, TX; Mountain 1: ID, MT, WY; Mountain 2: CO, NV, UT; Mountain 3: AZ, NM; Pacific: OR, WA.

Table A 1.13

Cash Labor Expenses Reported by the Economic Research Service: 1981-84

State	Cash Labor Expenses 1981-$mil	Per Dist	Cash Labor Expenses 1982-$mil	Per Dist	Cash Labor Expenses 1983-$mil	Per Dist	Cash Labor Expenses 1984-$mil	Per Dist
Alabama	80	1%	88	1%	84	1%	87	1%
Alaska	2		2		2		2	
Arizona	138	2%	156	2%	150	2%	155	2%
Arkansas	150	2%	165	2%	159	2%	163	2%
California	1566	21%	1819	21%	1749	22%	1800	22%
Colorado	114	2%	129	2%	124	2%	128	2%
Connecticut	39	1%	44	1%	43	1%	44	1%
Delaware	16		18		17		18	
Florida	421	6%	480	6%	462	6%	475	6%
Georgia	136	2%	150	2%	144	2%	148	2%
Hawaii	132	2%	146	2%	141	2%	145	2%
Idaho	126	2%	148	2%	142	2%	146	2%
Illinois	200	3%	226	3%	217	3%	223	3%
Indiana	132	2%	152	2%	146	2%	150	2%
Iowa	198	3%	222	3%	214	3%	220	3%
Kansas	136	2%	153	2%	148	2%	152	2%
Kentucky	141	2%	167	2%	160	2%	165	2%
Lousiana	98	1%	107	1%	103	1%	106	1%
Maine	40	1%	45	1%	43	1%	44	1%
Maryland	60	1%	69	1%	66	1%	68	1%
Massachusett	39	1%	43	1%	41	1%	43	1%
Michigan	159	2%	186	2%	179	2%	184	2%
Minnesota	179	2%	208	2%	200	2%	205	2%
Mississippi	125	2%	137	2%	132	2%	136	2%
Missouri	124	2%	141	2%	136	2%	140	2%
Montana	75	1%	85	1%	81	1%	84	1%

State	Cash Labor Expenses 1981-$mil	Per Dist	Cash Labor Expenses 1982-$mil	Per Dist	Cash Labor Expenses 1983-$mil	Per Dist	Cash Labor Expenses 1984-$mil	Per Dist
Nebraska	145	2%	168	2%	161	2%	166	2%
Nevada	19		20		20		20	
New Hampsh	12		14		13		14	
New Jersey	56	1%	62	1%	60	1%	62	1%
New Mexico	54	1%	61	1%	59	1%	61	1%
New York	212	3%	246	3%	237	3%	243	3%
North Caroli	222	3%	245	3%	236	3%	243	3%
North Dakot	70	1%	77	1%	74	1%	76	1%
Ohio	147	2%	166	2%	160	2%	164	2%
Oklahoma	86	1%	98	1%	95	1%	97	1%
Oregon	156	2%	180	2%	173	2%	178	2%
Pennsylvania	195	3%	224	3%	216	3%	222	3%
Rhode Island	5		6		5		6	
South Caroli	74	1%	80	1%	77	1%	79	1%
South Dakot	59	1%	67	1%	65	1%	67	1%
Tennessee	94	1%	109	1%	105	1%	108	1%
Texas	433	6%	481	6%	462	6%	475	6%
Utah	37	1%	42	1%	40		42	
Vermont	25		29		28		29	
Virginia	112	2%	127	2%	122	2%	126	2%
Washington	270	4%	313	4%	301	4%	310	4%
West Virgini	18		20		20		20	
Wisconsin	232	3%	279	3%	268	3%	276	3%
Wyoming	36		41		39		40	
United States	7394	100%	8441	100%	8115	100%	8350	100%

Source: FCRS data adjusted and published by the ERS in Economic Indicators of the Farm Sector

Cash labor expenses are wages paid to and payroll taxes paid for hired workers and expenses for labor contractors

Table A1.14

Cash Labor Expenses from the FCRS: 1982-86

Region	Cash Labor Expenses 1982-$000-	Per Dis	Cash Labor Expenses 1983-$000-	Per Dis	Cash Labor Expenses 1984-$000-	Per Dis	Cash Labor Expenses 1985-$000-	Per Dis	Cash Labor Expenses 1986-$000-	Per Di	Percent Change 1982-86	1985-86
Northeast	1,051,486	9%	1,061,920	10%	1,080,356	10%	1,058,931	11%	1,080,356	10%	3%	2%
Lake States	984,264	8%	1,057,013	10%	920,089	9%	797,636	8%	920,089	9%	-7%	15%
Cornbelt	949,378	8%	1,042,687	9%	1,221,531	12%	948,242	10%	1,221,531	12%	29%	29%
Northern Plai	620,055	5%	726,520	7%	631,125	6%	515,624	5%	631,124	6%	2%	22%
Applachian	1,011,251	9%	794,255	7%	632,821	7%	692,815	6%	632,821	6%	-37%	-9%
Southeast	973,731	8%	996,964	9%	818,043	9%	1,157,949	8%	818,043	8%	-16%	-29%
Delta States	575,453	5%	446,683	4%	547,205	4%	400,853	5%	547,205	5%	-5%	37%
Southern Plai	1,061,189	9%	1,255,006	11%	1,278,423	12%	784,220	12%	1,278,423	12%	20%	63%
Mountain	1,131,187	10%	948,308	9%	877,888	9%	795,454	8%	877,888	8%	-22%	10%
Pacific	3,228,679	28%	2,693,767	24%	2,464,852	24%	2,761,748	24%	2,464,852	24%	-24%	-11%
United States	11,586,673	100%	11,023,123	100%	10,472,333	100%	9,913,472	100%	10,472,332	100%	-10%	6%

Source: FCRS data published by NASS in Farm Production Expenditures

Cash labor expenses are wages paid to and payroll taxes paid for hired workers and expenses for labor contractors

Northeast: CT, DE, ME, MD, MA, NH, NJ, NY, PA, RI, VT; Lake: MI, MN, WI

Cornbelt: IL, IN, IA, MO, OH; Northern Plains: KS, NE, ND, SD; Applachian: KY, NC, TN, VA, WV

Southeast: AL, FL, GA, SC; Delta: AR, LA, MS; Southern Plains: OK, TX

Mountain: AZ, CO, ID, MT, NV, NM, UT, WY; Pacific: CA, OR, WA

Table A1.14a
BEA Farm Wages and Labor Expenses for 1982

State	BEA Total Farm Wages & Salaries 1982-$000-	Per Dist	BEA Hired Labor Expense 1982-$000-	Per Dist	BEA Contract Labor Expense 1982-$000-	Per Dist	BEA Other Farm Labor Income 1982-$000-	Per Dist
California	2,751,423	23%	2,121,933	20%	391,793	34%	264,907	39%
Florida	689,492	6%	581,252	6%	219,938	19%	55,098	8%
Texas	650,841	6%	640,915	6%	106,604	9%	15,782	2%
Washington	394,607	3%	383,563	4%	30,208	3%	9,929	1%
North Carolina	344,539	3%	330,040	3%	25,759	2%	9,483	1%
Iowa	344,474	3%	300,653	3%	8,647	1%	7,624	1%
Wisconsin	338,066	3%	316,447	3%	9,683	1%	10,230	2%
Illinois	314,150	3%	304,216	3%	7,323	1%	28,505	4%
New York	297,498	3%	288,269	3%	13,093	1%	24,493	4%
Pennsylvania	262,271	2%	254,785	2%	13,237	1%	6,895	1%
Top Ten States	6,387,361	55%	5,522,073	53%	826,285	71%	432,946	65%
United States	11,714,000	100%	10,419,364	100%	1,162,257	100%	671,000	100%

Source: Bureau of Economic Analysis, U.S. Department of Commerce; Farm Income and Expenditures series

BEA state and county estimates are based on FCRS data from ERS; ERS lumps labor expenditures and contract labor together, while BEA separates them

Hired farm labor expense is similar to the COA labor expenditures concept; it include cash wages paid to all wage and salary workers, employer payroll taxes and the value of any fringe benefits offered to workers; excluding contract labor ex

Farm wages and salaries include cash wages paid to hired workers, in-kind benefits for hired workers

and the salaries received by officers of corporate farms; they exclude contract labor expenses

Table A 1.15

BEA Estimates of Farm Labor Expenses: 1979-1984

State	Hired Farm Labor Expens 1979-$000-	Per Dist	Hired Farm Labor Expens 1980-$000-	Per Dist	Hired Farm Labor Expens 1981-$000-	Per Dist	Hired Farm Labor Expens 1982-$000-	Per Dist	Hired Farm Labor Expens 1983-$000-	Per Dist	Hired Farm Labor Expens 1984-$000-	Per Dist
Alabama	96,260	1%	104,195	1%	104,559	1%	114,676	1%	110,146	1%	113,374	1%
Alaska	2,350		2,740		2,554		2,813		2,699		2,777	
Arizona	134,464	2%	146,316	2%	154,121	2%	174,764	2%	168,057	2%	172,826	2%
Arkansas	156,681	2%	198,117	2%	186,169	2%	205,555	2%	197,505	2%	203,251	2%
California	1,721,988	20%	1,873,217	20%	1,829,397	20%	2,121,933	20%	2,040,390	20%	2,098,687	20%
Colorado	115,458	1%	147,263	2%	140,358	2%	159,369	2%	153,166	2%	157,534	2%
Connecticut	49,851	1%	54,534	1%	53,851	1%	61,334	1%	58,909	1%	60,607	1%
Delaware	18,238		19,036		19,746		22,321		21,451		22,065	
Florida	483,474	6%	510,376	6%	510,161	6%	581,252	6%	559,091	6%	574,994	6%
Georgia	161,415	2%	191,250	2%	181,781	2%	200,487	2%	192,623	2%	198,214	2%
Hawaii	157,431	2%	185,974	2%	170,533	2%	189,852	2%	182,473	2%	187,774	2%
Idaho	136,337	2%	134,699	1%	142,389	2%	165,767	2%	159,371	2%	163,907	2%
Illinois	282,683	3%	248,277	3%	269,031	3%	304,216	3%	292,227	3%	300,710	3%
Indiana	120,381	1%	146,181	2%	142,279	2%	164,180	2%	157,831	2%	162,357	2%
Iowa	284,948	3%	235,007	3%	267,018	3%	300,653	3%	288,851	3%	297,163	3%
Kansas	189,384	2%	208,482	2%	195,718	2%	220,894	2%	212,171	2%	218,353	2%
Kentucky	109,117	1%	107,811	1%	121,791	1%	143,808	1%	138,321	1%	142,257	1%
Lousiana	117,973	1%	133,617	1%	129,958	1%	142,823	1%	137,189	1%	141,201	1%
Maine	50,579	1%	52,153	1%	52,198	1%	58,340	1%	56,031	1%	57,682	1%
Maryland	71,865	1%	72,311	1%	74,286	1%	85,513	1%	82,178	1%	84,526	1%
Massachusetts	57,074	1%	62,486	1%	60,688	1%	67,110	1%	64,450	1%	66,312	1%
Michigan	178,832	2%	228,647	2%	202,354	2%	236,042	2%	226,829	2%	233,353	2%
Minnesota	184,066	2%	181,065	2%	196,460	2%	227,142	2%	218,343	2%	224,570	2%
Mississippi	139,463	2%	175,231	2%	162,400	2%	178,532	2%	171,480	2%	176,519	2%
Missouri	147,803	2%	161,882	2%	161,666	2%	184,185	2%	176,985	2%	182,036	2%
Montana	115,018	1%	125,793	1%	117,506	1%	133,374	1%	128,093	1%	131,790	1%
Nebraska	173,262	2%	190,172	2%	183,771	2%	212,098	2%	203,806	2%	209,670	2%
Nevada	23,603		25,182		25,396		28,031		26,933		27,707	
New Hampshi	14,287		15,680		15,536		17,294		16,620		17,100	
New Jersey	70,211	1%	79,566	1%	76,697	1%	85,450	1%	82,108	1%	84,479	1%
New Mexico	87,927	1%	85,720	1%	83,896	1%	94,823	1%	91,110	1%	93,728	1%

State	Hired Farm Labor Expen 1979-$000	Per Dist	Hired Farm Labor Expen 1980-$000	Per Dist	Hired Farm Labor Expen 1981-$000	Per Dist	Hired Farm Labor Expen 1982-$000	Per Dist	Hired Farm Labor Expen 1983-$000	Per Dist	Hired Farm Labor Expen 1983-$000	Per Dist
New York	232,081	3%	249,941	3%	248,164	3%	288,269	3%	277,054	3%	285,011	3%
North Carolina	307,946	4%	273,435	3%	297,292	3%	330,040	3%	317,087	3%	326,292	3%
North Dakota	86,297	1%	92,874	1%	93,150	1%	103,240	1%	99,191	1%	102,052	1%
Ohio	169,422	2%	168,920	2%	177,182	2%	200,533	2%	192,711	2%	198,270	2%
Oklahoma	108,794	1%	113,656	1%	111,938	1%	127,483	1%	122,517	1%	126,046	1%
Oregon	149,262	2%	164,844	2%	169,555	2%	195,241	2%	187,701	2%	193,055	2%
Pennsylvania	190,527	2%	230,157	2%	221,914	2%	254,785	2%	244,901	2%	251,917	2%
Rhode Island	4,557		5,095		5,276		6,002		5,770		5,937	
South Carolina	109,759	1%	135,691	1%	120,017	1%	130,323	1%	125,142	1%	128,811	1%
South Dakota	78,296	1%	77,493	1%	78,452	1%	89,581	1%	86,078	1%	88,552	1%
Tennessee	107,671	1%	118,990	1%	113,911	1%	132,315	1%	127,170	1%	130,831	1%
Texas	547,277	6%	571,352	6%	576,026	6%	640,915	6%	615,838	6%	633,667	6%
Utah	37,712		41,937		41,808		47,420		45,585		46,900	
Vermont	30,363		33,349		32,472		37,699		36,221		37,256	
Virginia	122,336	1%	147,969	2%	140,881	2%	159,734	2%	153,503	2%	157,903	2%
Washington	312,552	4%	352,250	4%	330,800	4%	383,563	4%	368,632	4%	379,304	4%
West Virginia	22,000		18,596		21,332		24,242		23,302		23,971	
Wisconsin	264,808	3%	271,991	3%	264,518	3%	316,447	3%	304,162	3%	312,898	3%
Wyoming	56,986	1%	60,284	1%	58,447	1%	66,901	1%	64,255	1%	66,080	1%
United States	8,591,069	100%	9,231,804	100%	9,137,403	100%	10,419,364	100%	10,014,257	100%	10,302,276	100%

Source: Bureau of Economic Analysis, U.S. Department of Commerce; Farm Income and Expenditures series

BEA state and county estimates are based on FCRS data from ERS; ERS lumps labor expenditures and contract labor together, while BEA separates them

Hired farm labor expense is similar to the COA labor expenditures concept; it include cash wages paid to all wage and salary workers, employer payroll taxes and the value of any fringe benefits offered to workers; it excludes contract labor expenses

Farm wages and salaries include cash wages paid to hired workers, in-kind benefits for hired workers and the salaries received by officers of corporate farms; it excludes contract labor expenses

Table A 1.16

Farmers in the 1980 Census of Population

State	All Farmers							Female Farmers						
	Tot Farmers Occ 473	Per Dist	White	Total Minority	Per Dist	Black	Hispanic	Total Farmers	Per Dist	White	Total Minority	Per Dist	Black	Hispanic
Alabama	14,120	1%	13,335	785	2%	683	63	1,476	1%	1,403	73	2%	59	8
Alaska	191		182	9			7	22		22			-	-
Arizona	3,134	2%	2,684	450	1%	17	290	316	3%	276	40	1%	-	23
Arkansas	22,173	3%	21,373	800	2%	583	114	2,866	4%	2,832	34	1%	14	20
California	37,509	1%	31,646	5,863	16%	188	2,938	4,710	1%	3,844	866	21%	63	380
Colorado	14,687		14,153	534	1%	10	420	1,509		1,481	28	1%	6	16
Connecticut	2,246		2,224	22		6	16	396		396	-		-	-
Delaware	1,680		1,655	25		19	-	316		312	4		4	-
Florida	13,546	1%	11,868	1,678	5%	948	694	2,069	2%	1,927	142	3%	83	59
Georgia	22,155	2%	20,568	1,587	4%	1,432	127	2,151	2%	2,007	144	3%	125	12
Hawaii	1,673		419	1,254	3%	-	45	400		93	307	7%	-	18
Idaho	14,464	1%	13,875	589	2%	-	195	907	1%	876	31	1%	-	19
Illinois	61,664	6%	61,416	248	1%	69	117	4,761	4%	4,687	74	2%	26	27
Indiana	37,144	3%	36,999	145		24	83	2,982	3%	2,962	20		-	4
Iowa	86,591	8%	86,433	158		10	109	6,769	6%	6,746	23	1%	-	17
Kansas														
Kentucky	35,825	3%	35,160	665	2%	445	193	2,553	2%	2,524	29	1%	17	6
Louisiana	11,636	1%	10,538	1,098	3%	1,001	64	681	1%	630	51	1%	39	5
Maine	3,757		3,743	14		-	4	495		492	3		-	-
Maryland	8,495	1%	8,084	411	1%	387	1	1,353	1%	1,313	40	1%	35	-
Massachusetts	2,599		2,569	30		5	6	455		449	6		-	-
Michigan	26,647	2%	26,451	196	1%	113	39	2,696	2%	2,662	34	1%	17	8
Minnesota	69,692	6%	69,552	140		5	67	6,159	6%	6,126	33	1%	-	26
Mississippi	12,987	1%	11,556	1,431	4%	1,378	34	903	1%	821	82	2%	71	6
Missouri	53,875	5%	53,580	295	1%	71	150	5,526	5%	5,488	38	1%	-	22
Montana	14,126	1%	13,831	295	1%	7	43	1,362	1%	1,344	18		-	10
Nebraska	47,951	4%	47,788	163		12	71	3,027	3%	3,006	21	1%	-	11
Nevada	1,237		1,165	72		-	43	152		138	14		-	-
New Hamps	1,475		1,460	15		-	-	311		308	3		-	-
New Jersey	4,122		4,045	77		48	18	563	1%	544	19		5	12
New Mexico	5,023		4,028	995	3%	4	917	378		297	81	2%	-	76

All Farmers

State	Total Farmers	Per Dist	White	Total Minority	Per Dist	Black	Hispanic
New York	29,176	3%	28,895	281	1%	84	127
North Carol	36,927	3%	32,928	3,999	11%	3,393	146
North Dako	30,588	3%	30,450	138		11	9
Ohio	39,955	4%	39,712	243	1%	75	135
Oklahoma	28,219	3%	27,199	1,020	3%	201	92
Oregon	13,244	1%	12,827	417	1%	14	103
Pennsylvani	31,975	3%	31,783	192	1%	29	120
Rhode Islan	383		377	6		–	6
South Carol	9,768	1%	7,791	1,977	5%	1,878	84
South Dako	33,214	3%	32,812	402	1%	6	19
Tennessee	24,621	2%	23,884	737	2%	606	100
Texas	70,094	6%	65,780	4,314	12%	844	3,263
Utah	4,690		4,577	113		–	3
Vermont	4,733		4,708	25		–	11
Virginia	19,472	2%	18,218	1,254	3%	1,157	80
Washington	15,713	1%	15,185	528	1%	25	190
West Virgin	3,889		3,859	30		15	15
Wisconsin	67,478	6%	67,325	153		11	78
Wyoming	4,497		4,332	165		–	71
United State	1,101,060	100%	1,065,022	36,038	100%	15,814	11,520

Female Farmers

State	Total Farmers	Per Dist	White	Total Minority	Per Dist	Black	Hispanic
New York	3,602	3%	3,545	57	1%	17	33
North Carol	3,461	3%	2,917	544	13%	483	11
North Dako	1,903	2%	1,899	4		–	–
Ohio	3,820	4%	3,760	60	1%	15	31
Oklahoma	2,724	3%	2,612	112	3%	12	15
Oregon	1,966	2%	1,914	52	1%	–	25
Pennsylvani	3,738	3%	3,711	27	1%	3	8
Rhode Islan	70		70	–		–	–
South Carol	1,051	1%	700	351	8%	322	14
South Dako	2,649	2%	2,602	47	1%	25	10
Tennessee	1,786	2%	1,748	38	1%	25	13
Texas	6,063	6%	5,724	339	8%	53	247
Utah	162		154	8		–	–
Vermont	759	1%	754	5		–	2
Virginia	1,958	2%	1,871	87	2%	71	11
Washington	1,922	2%	1,818	104	3%	6	24
West Virgin	472		472	–		–	–
Wisconsin	11,966	11%	11,915	51	1%	3	20
Wyoming	429		429	–		–	–
United State	108,765	100%	104,621	4,144	100%	1,574	1,249

Source: Census of Population, 1980; employment data is based on a 19 percent sample of households

Occupation is based on the responding individual's chief job activity during the last week in March, 1980

Farmers includes all farm operators and managers except hort specialty farmers

Table A 1.17

Farmworkers in the 1980 Census of Population

| | All Farmworkers | | | | | | | | Female Farmworkers | | | | | | |
State	All Farmworke Occ 479	Per Dist	White	Total Minority	Per Dist	Black	Hispanic	Per Dist	Female Farmworkers	Per Dist	White	Total Minority	Per Dist	Black	Hispanic
Alabama	10,147	1%	6,744	3,403	1%	3,203	137		2,147	1%	1,668	479	1%	437	24
Alaska	3,881		342	46		-	15		63		44	19		-	14
Arizona	14,282	2%	4,547	9,735	3%	355	8,293	4%	2,138	1%	415	1,723	2%	26	1,569
Arkansas	20,120	2%	15,137	4,983	2%	4,528	297		2,916	2%	2,523	393	2%	302	57
California	150,116	17%	34,493	115,623	38%	2,147	106,115	56%	33,407	18%	6,098	27,309	18%	363	25,122
Colorado	11,270	1%	9,095	2,175	1%	23	1,962	1%	1,827	1%	1,662	165	1%	-	139
Connecticut	3,057		2,714	343		47	280		886		798	88		15	66
Delaware	1,901		1,390	511		441	53		638		463	175		144	17
Florida	51,824	6%	19,142	36,682	12%	20,481	11,634	6%	14,481	8%	5,616	8,865	8%	5,708	2,949
Georgia	21,969	3%	12,169	9,800	3%	9,292	410		4,695	2%	2,698	1,997	2%	1,872	102
Hawaii	5,855	1%	812	5,043	2%	18	487		1,570	1%	77	1,493	1%	-	120
Idaho	12,982	1%	9,916	3,066	1%	-	2,825	1%	2,054	1%	1,781	273	1%	-	261
Illinois	25,522	3%	24,439	1,083		283	686		5,156	3%	4,971	185	3%	43	120
Indiana	16,683	2%	16,362	321		126	153		3,736	2%	3,647	89	2%	25	50
Iowa	30,022	3%	29,831	191		22	98		8,354	4%	8,319	35	4%	4	5
Kansas															
Kentucky	17,515	2%	16,250	1,265	2%	1,002	236		2,414	1%	2,292	122	1%	93	29
Louisiana	12,228	1%	6,061	6,167		5,851	227	2%	1,331	1%	660	671	1%	624	39
Maine	4,271		4,170	101		14	24		892		870	22		-	8
Maryland	7,259	1%	5,787	1,472		1,287	113		1,801	1%	1,517	284	1%	260	24
Massachusetts	4,033		3,651	382		11	351		1,200	1%	1,114	86	1%	5	74
Michigan	18,618	2%	17,337	1,281		282	836		4,789	3%	4,416	373	3%	65	282
Minnesota	26,111	3%	25,848	263		9	137		7,754	4%	7,686	68	4%	-	33
Mississippi	16,810	2%	5,549	11,261	4%	10,911	266	4%	2,037	1%	798	1,239	1%	1,207	25
Missouri	21,014	2%	20,065	949		729	125		4,690	2%	4,570	120	2%	85	22
Montana	10,474	1%	9,934	540		5	83		2,003	1%	1,951	52	1%	-	22
Nebraska	18,868	2%	18,511	357		27	210		3,685	2%	3,607	78	2%	16	40
Nevada	1,968		1,367	601		37	442		213		188	25		5	5
New Hampshire	1,562		1,524	38		10	21		360		346	14		7	7
New Jersey	5,144	1%	4,252	892		392	476		1,187	1%	1,070	117	1%	72	45
New Mexico	6,556	1%	2,650	3,906	1%	47	3,489	2%	684	1%	299	385	1%	2	318

All Farmworkers

State	All Farmworkers	Per Dist	White	Total Minority	Per Dist	Black	Hispanic	Per Dist
New York	27,800	3%	26,046	1,754	1%	902	628	
North Caro	29,407	3%	14,708	14,699	5%	13,433	416	
North Dak	8,127	1%	7,931	196		7	89	
Ohio	19,591	2%	18,668	923	1%	258	611	
Oklahoma	11,637	1%	10,031	1,606	1%	362	496	
Oregon	16,051	2%	12,718	3,333	1%	59	2,878	2%
Pennsylvan	22,526	3%	21,675	851		250	562	
Rhode Islan	504		499	5		–	5	
South Caro	9,120	1%	3,259	5,861	2%	5,503	266	
South Dak	10,975	1%	10,503	472		10	26	
Tennessee	12,787	1%	10,703	2,084	1%	1,886	159	
Texas	65,276	7%	26,778	38,498	13%	3,990	34,267	18%
Utah	4,421	1%	3,991	430		–	322	
Vermont	4,029		4,009	20		–	11	
Virginia	15,139	2%	10,847	4,292	1%	4,059	158	
Washingto	23,381	3%	14,967	8,414	3%	155	7,344	4%
West Virgi	3,571		3,420	151		99	39	
Wisconsin	34,688	4%	34,278	410		47	214	
Wyoming	3,692		3,333	359		–	291	
United Stat	874,784	100%	568,453	306,838	100%	92,600	189,263	100%

Female Farmworkers

State	Female Farmworkers	Per Dist	White	Total Minority	Black	Hispanic
New York	6,198	3%	5,927	271	132	85
North Caro	8,021	4%	3,625	4,396	4,096	49
North Dak	1,359	1%	1,319	40	–	33
Ohio	5,363	3%	5,077	286	67	185
Oklahoma	2,085	1%	1,881	204	78	21
Oregon	3,405	2%	2,930	475	6	345
Pennsylvan	5,717	3%	5,637	80	43	35
Rhode Islan	89		84	5	–	5
South Caro	2,044	1%	603	1,441	1,358	68
South Dak	2,409	1%	2,382	27	–	–
Tennessee	1,662	1%	1,474	188	174	7
Texas	10,357	5%	4,290	6,067	514	5,532
Utah	477		442	35	–	24
Vermont	753		750	3	–	3
Virginia	2,708	1%	1,946	762	709	42
Washingto	5,466	3%	3,565	1,901	8	1,584
West Virgi	695		690	5	–	5
Wisconsin	11,261	6%	11,167	94	19	39
Wyoming	482		479	3	–	–
United Stat	189,659	100%	126,432	63,227	18,579	39,650

Source: Census of Population, 1980

Occupation is based on the responding individual's chief job activity during the last week in March, 1980

Farmers includes all farm operators and managers except hort specialty farmers

Table A1.18
Agricultural Employment and Wages Covered by Unemployment Insurance in 1985

State	Fips Cod	Crop Employers	Per Dist	Ave Annual Employment	Per Dist	Annual Wages($000)	Per Dist	Livestock Employers	Per Dist	Ave Annual Employmen	Per Dist	Annual Wages($000)	Per Dist
Alabama	1	143	1%	2,746	1%	28,646	1%	98	1%	2,794	2%	33,708	2%
Alaska	2	7		89		1,104		9		42		750	
Arizona	4	535	2%	11,710	2%	139,881	3%	170	1%	2,501	2%	35,656	2%
Arkansas	5	522	1%	4,179	1%	48,878	1%	121	1%	3,578	3%	48,874	3%
California	6	16,678	47%	199,427	40%	2,190,926	42%	4,425	37%	29,201	22%	392,532	23%
Colorado	8	258	1%	4,105	1%	46,618	1%	210	2%	2,558	2%	38,160	2%
Connecticu	9	113		4,535	1%	44,692	1%	42		1,291	1%	20,444	1%
Delaware	10	44		637		6,196		12		401		5,810	
Florida	12	1,766	5%	55,350	11%	600,913	11%	433	4%	7,404	6%	94,150	5%
Georgia	13	301	1%	5,927	1%	53,602	1%	165	1%	3,477	3%	43,181	2%
Hawaii	15	113		8,186	2%	129,468	2%	58		948		14,315	1%
Idaho	16	334	1%	6,073	1%	57,584	1%	156	1%	1,916	1%	26,247	2%
Illinois	17	522	1%	12,051	2%	121,931	2%	138	1%	986	1%	12,628	1%
Indiana	18	299	1%	6,809	1%	66,259	1%	114	1%	2,278	2%	30,000	2%
Iowa	19	206	1%	2,025		22,432		145	1%	1,363	1%	18,223	1%
Kansas	20	188	1%	1,586		22,319		232	2%	3,304	2%	55,629	3%
Kentucky	21	135		1,365		13,179		98	1%	2,559	2%	37,449	2%
Lousiana	22	541	2%	5,113	1%	52,418	1%	72	1%	848	1%	10,151	1%
Maine	23	84		1,071		10,627		29		912		8,870	1%
Maryland	24	141		2,106		22,269		64	1%	1,603	1%	25,611	1%
Massachuse	25	181	1%	3,134	1%	39,910	1%	55		715		9,327	1%
Michigan	26	625	2%	10,362	2%	103,411	2%	85	1%	1,284	1%	15,378	1%
Minnesota	27	463	1%	4,143	1%	46,855	1%	258	2%	2,827	2%	33,259	2%
Mississippi	28	522	1%	6,147	1%	53,761	1%	87	1%	2,586	2%	31,684	2%
Missouri	29	407	1%	3,650	1%	37,513	1%	147	1%	1,529	1%	18,015	1%
Montana	30	90		632		7,296		157	1%	1,298	1%	14,007	1%
Nebraska	31	174		1,629		20,419		248	2%	2,560	2%	38,171	2%
Nevada	32	32		458		6,033		84	1%	999	1%	11,324	1%
New Hamp	33	35		641		7,270		12		282		5,235	
New Jersey	34	652	2%	7,185	1%	80,382	2%	80	1%	694	1%	9,197	1%
New Mexic	35	121		2,456		25,505		145	1%	2,199	2%	26,081	2%

Fips Cod	State	Crop Employers	Per Dist	Ave Annual Employment	Per Dist	Annual Wages($000)	Per Dist	Livestock Employers	Per Dist	Ave Annual Employment	Per Dist	Annual Wages($000)	Per Dist
36	New York	812	2%	9,960	2%	111,017	2%	271	2%	3,406	3%	42,535	2%
37	North Caro	429	1%	6,367	1%	53,489	1%	160	1%	4,807	4%	65,937	4%
38	North Dakc	303	1%	1,213		13,811		31		264		3,220	
39	Ohio	442	1%	8,132	2%	89,751	2%	123	1%	1,870	1%	20,381	1%
40	Oklahoma	93		1,839		22,042		160	1%	1,674	1%	22,714	1%
41	Oregon	667	2%	13,679	3%	120,592	2%	146	1%	1,685	1%	21,006	1%
42	Pennsylvan	471	1%	11,113	2%	123,185	2%	160	1%	3,220	2%	38,386	2%
43	Puerto Ric	2,431	7%	11,242	2%	31,366	1%	825	7%	4,395	3%	22,244	1%
44	Rhode Islan	93		681		8,321		36		265		2,349	
45	South Caro	200	1%	3,865	1%	30,890	1%	63	1%	1,493	1%	17,107	1%
46	South Dakc	32		250		3,006		67	1%	512		6,848	
47	Tennessee	194	1%	2,855	1%	30,360	1%	70	1%	642		7,110	
48	Texas	737	2%	17,121	3%	178,669	3%	979	8%	12,574	9%	187,053	11%
49	Utah	38		872		8,730		46		639		7,198	
50	Vermont	33		373		3,284		28		214		2,551	
51	Virginia	196	1%	3,540	1%	32,398	1%	179	1%	2,420	2%	29,110	2%
52	Virgin Islar	6		7		48		10		43		396	
53	Washington	1,271	4%	26,831	5%	202,273	4%	172	1%	2,862	2%	36,749	2%
54	West Virgi	47		848		6,565		32		164		1,750	
55	Wisconsin	410	1%	5,813	1%	68,376	1%	154	1%	1,797	1%	21,331	1%
56	Wyoming	24		171		1,783		90	1%	1,074	1%	12,312	1%
57	United Stat	35,155	100%	502,295	100%	5,248,257	100%	11,944	100%	132,953	100%	1,732,349	100%

Source: ES-202 Employment and Wages Program; Bureau of Labor Statistics, U.S. Dept of Labor

Most states have Federal Unemployment Tax Act (FUTA) coverage of farm employers; farm employers who pay quarterly cash wages of $20,000 or more or employ 10 or more workers for at least one day in each of 20 different weeks must cover their farmworkers (20-10 rule). CA, ME, MN, RI, TX, and WA have more inclusive farmworker coverage.

In 1976, DOL estimated that 40 percent of all farmworkers were employed on farms required by federal law to provide UI coverage

Changes in agriculture have raised coverage levels; in California, the 30% of UI-covered crop and livestock employers who each paid $ or more in farm wages paid 81 percent of all farm wages and accounted for 90 % of all farm jobs

Employers are farms or their subunits which employ 50 or more workers; this column is the number of ers reporting each quarter divide

Ave annual employment is the employer-reported count of all workers on the payroll for the pay period which includes the 12th day of the month summed over 12 months and divided by 12; this column is the average of four quarterly reports.

Table A1.18
Agricultural Employment and Wages Covered by Unemployment Insurance in 1985(Continued)

State	Ag Service Employers	Per Dist	Ave Annual Employmen	Per Dist	Annual Wages($000)	Per Dist	All Ag Employers	Per Dist	Ave Annual Employmen	Per Dist	Annual Wages($000)	Per Dist
Alabama	783	1%	5,178	1%	55,146	1%	917	1%	10,717	1%	117,499	1%
Alaska	106		341		6,519		116		472		8,374	
Arizona	1,407	2%	15,297	3%	146,193	2%	1,711	2%	29,507	3%	321,730	2%
Arkansas	695	1%	3,353	1%	36,283	1%	946	1%	11,110	1%	134,035	1%
California	9,829	14%	147,789	28%	1,628,930	26%	18,423	20%	376,417	32%	4,212,388	32%
Colorado	1,180	2%	8,167	2%	102,771	2%	1,455	2%	14,830	1%	187,549	1%
Connecticut	1,306	2%	6,372	1%	86,470	1%	1,376	2%	12,198	1%	151,606	1%
Delaware					0		23		1,038		12,006	
Florida	5,685	8%	56,608	11%	565,507	9%	6,559	7%	119,362	10%	1,260,571	10%
Georgia	1,483	2%	10,781	2%	130,973	2%	1,723	2%	20,185	2%	227,757	2%
Hawaii	209		1,895		25,529		295		11,028	1%	169,312	1%
Idaho	391	1%	2,761	1%	29,442		630	1%	10,750	1%	113,273	1%
Illinois	2,371	3%	15,492	3%	215,347	3%	2,640	3%	28,529	2%	349,906	3%
Indiana	1,229	2%	7,890	1%	78,419	1%	1,417	2%	16,977	1%	174,678	1%
Iowa	1,122	2%	4,224	1%	50,772	1%	1,318	1%	7,612	1%	91,426	1%
Kansas	816	1%	3,611	1%	42,950	1%	1,094	1%	8,501	1%	120,899	1%
Kentucky	930	1%	7,008	1%	87,550	1%	1,062	1%	10,932	1%	138,178	1%
Lousiana	954	1%	5,867	1%	66,952	1%	1,161	1%	11,829	1%	129,521	1%
Maine	342		1,766		19,359		392		3,750		38,856	
Maryland	1,331	2%	9,906	2%	125,173	2%	1,431	2%	13,614	1%	173,054	1%
Massachuset	1,584	2%	10,623	2%	156,509	3%	1,683	2%	14,472	1%	205,746	2%
Michigan	1,885	3%	11,905	2%	149,823	2%	2,126	2%	23,550	2%	268,612	2%
Minnesota	1,227	2%	8,184	2%	92,656	1%	1,600	2%	15,153	1%	172,769	1%
Mississippi	827	1%	3,805	1%	36,197	1%	1,044	1%	12,538	1%	121,642	1%
Missouri	1,330	2%	7,613	1%	86,486	1%	1,578	2%	12,792	1%	142,015	1%
Montana	321		909		9,418		500	1%	2,838		30,720	
Nebraska	631	1%	3,043	1%	32,060	1%	923	1%	7,232	1%	90,651	1%
Nevada	325		1,592		21,348		416		3,048		38,705	
New Hampsh	354	1%	1,606		19,131		374		2,529		31,636	
New Jersey	2,488	4%	13,351	2%	184,131	3%	2,731	3%	21,230	2%	273,710	2%
New Mexico	411	1%	2,932	1%	25,665		586	1%	7,587	1%	77,251	1%

State	Ag Service Employers	Per Dist	Ave Annual Employmen	Per Dist	Annual Wages($000)	Per Dist	All Ag Employers	Per Dist	Ave Annual Employmen	Per Dist	Annual Wages($000')	Per Dist
New York	4,548	7%	21,324	4%	298,416	5%	5,021	6%	34,690	3%	451,968	3%
North Caroli	1,596	2%	9,979	2%	110,251	2%	1,863	2%	21,153	2%	229,677	2%
North Dakot	232		776		9,498		338		2,253		26,528	
Ohio	2,841	4%	15,575	3%	202,977	3%	3,075	3%	25,576	2%	313,110	2%
Oklahoma	686	1%	4,002	1%	46,840	1%	869	1%	7,515	1%	91,595	1%
Oregon	943	1%	4,333	1%	49,477	1%	1,256	1%	19,697	2%	191,075	1%
Pennsylvani	2,782	4%	16,090	3%	200,548	3%	3,060	3%	30,423	3%	362,119	3%
Puerto Rico	118		531		2,883		1,551	2%	16,168	1%	56,493	
Rhode Island	348	1%	1,426	1%	17,414		408		2,372		28,085	
South Caroli	796	1%	5,057	1%	52,995	1%	909	1%	10,415	1%	100,993	1%
South Dakot	255		819		10,061		329		1,581		19,915	
Tennessee	941	1%	6,144	1%	64,977	1%	1,059	1%	9,641	1%	102,447	1%
Texas	4,470	6%	35,633	7%	432,092	7%	5,633	6%	65,327	6%	797,814	6%
Utah	336		2,187		24,532		391		3,698		40,460	
Vermont	212		1,196		14,975		248		1,782		20,810	
Virginia	1,333	2%	9,992	2%	119,478	2%	1,561	2%	15,952	1%	180,986	1%
Virgin Island	16		65		615		28		115		1,059	
Washington	1,814	3%	9,139	2%	95,463	2%	2,304	3%	38,831	3%	334,485	3%
West Virgini	235		1,513	1%	15,160		279		2,525		23,475	
Wisconsin	1,133	2%	7,991	2%	115,091	2%	1,390	2%	15,600	1%	204,797	2%
Wyoming	160		544		5,540		256		1,788		19,635	
United State	69,594	100%	535,553	100%	6,202,993	100%	90,327	100%	1,170,801	100%	13,183,599	100%

Source: ES-202 Employment and Wages Program; Bureau of Labor Statistics, U.S. Dept of Labor

Most states have Federal Unemployment Tax Act (FUTA) coverage of farm employers; farm employers who pay quarterly cash wages of $20,000 or more or employ 10 or more workers for at least one day in each of 20 different weeks must cover their farmworkers (20-10 rule). CA, ME, MN, RI, TX, and WA have more inclusive farmworker coverage.

In 1976, DOL estimated that 40 percent of all farmworkers were employed on farms required by federal law to provide UI coverage

Changes in agriculture have raised coverage levels; in California, the 30% of UI-covered crop and livestock employers who each paid $50,0 or more in farm wages paid 81 percent of all farm wages and accounted for 90 % of all farm jobs

Employers are farms or their subunits which employ 50 or more workers; this column is the number of ers reporting each quarter divided by

Ave annual employment is the employer-reported count of all workers on the payroll for the pay period which includes the 12th day of the month summed over 12 months and divided by 12; this column is the average of four quarterly reports.

Table A1.19

UI-Covered Employment and Wages Covered in Fruits and Vegetables: 1985

Fips	State	Vegetable Employer	Per Dist	Ave Annual Employment	Per Dist	Annual Wages($000)	Per Dist	Fruit&Nut Employer	Per Dist	Ave Annual Employment	Per Dist	Annual Wages($000)	Per Dist
1	Alabama												
2	Alaska	5		37		210		5		34		237	
4	Arizona	56	2%	3,014	4%	39,320	4%	49		2,398	2%	23,481	2%
5	Arkansas					0		6		40		388	
6	California	1,146	39%	36,690	43%	496,447	54%	9,162	71%	82,705	57%	703,818	57%
8	Colorado	35	1%	679	1%	7,126	1%	10		146		906	
9	Connecticut	13		91		1,095		10		202		2,342	
10	Delaware	3		130		1,302		3		84		797	
12	Florida	410	14%	22,286	26%	180,559	20%	554	4%	10,000	7%	112,303	9%
13	Georgia	18	1%	235		1,671		67	1%	1,526	1%	12,019	1%
15	Hawaii	13		178		1,581		24		2,887	2%	42,636	3%
16	Idaho	4		78		395		13		806	1%	5,800	
17	Illinois	15	1%	428		4,533		22		409		2,728	
18	Indiana	17	1%	209		1,875		17		322		2,672	
19	Iowa	3		14		74						0	
20	Kansas					0		2		9		62	
21	Kentucky					0						0	
22	Lousiana					0		5		87		743	
23	Maine	7		99		820		29		326		4,255	
24	Maryland	4		78		816		15		233		1,962	
25	Massachuse	21	1%	268		2,440		60		987	1%	14,975	1%
26	Michigan	46	2%	717	1%	6,929	1%	155	1%	3,123	2%	24,645	2%
27	Minnesota	25	1%	523	1%	5,229	1%	13		307		1,638	
28	Mississippi					0		4		53		563	
29	Missouri	5		41		345		32		551		3,288	
30	Montana					0		5		7		66	
31	Nebraska					0						0	
32	Nevada					0						0	
33	New Hamps					0		8		281		2,190	

State	Fips	Vegetable Employer	Per Dist	Ave Annual Employment	Per Dist	Annual Wages($000)	Per Dist	Fruit&Nut Employer	Per Dist	Ave Annual Employment	Per Dist	Annual Wages($000)	Per Dist
New Jersey	34	337	11%	3,119	4%	28,243	3%	91	1%	1,400	1%	15,935	1%
New Mexico	35	20	1%	368		2,798		20		681		8,186	1%
New York	36	224	8%	2,588	3%	27,954	3%	226	2%	3,338	2%	31,554	3%
North Carol	37	21	1%	318		1,847		23		398		2,430	
North Dako	38					0						0	
Ohio	39	45	2%	1,178	1%	10,924	1%	23	1%	439		3,044	
Oklahoma	40					0		6		70		711	
Oregon	41	103	3%	1,626	2%	11,316	1%	187	1%	4,409	3%	29,056	2%
Pennsylvani	42	24	1%	359		3,068		87	1%	1,727	1%	14,736	1%
Puerto Rico	43	8		455	1%	2,597		1,020	8%	3,631	3%	7,479	1%
Rhode Islan	44	11		42		368		13		36		218	
South Carol	45	22	1%	418		2,991		53	1%	1,227	1%	7,845	1%
South Dako	46					0						0	
Tennessee	47	16	1%	198		991						0	
Texas	48	78	3%	5,753	7%	43,055	5%	48		1,197	1%	12,439	1%
Utah	49	3		44		697		7		166		1,211	
Vermont	50					0		15		241		2,060	
Virginia	51	13		493	1%	2,957	1%	55		1,179	1%	10,175	1%
Virgin Islan	52	4		4		23						0	
Washington	53	129	4%	2,091	2%	17,038	2%	638	5%	15,690	11%	97,929	8%
West Virgir	54					0		29		664		4,730	
Wisconsin	55	60	2%	1,151	1%	12,550	1%	76	1%	857	1%	14,539	1%
Wyoming	56					0						0	
United Stat	57	2,960	100%	85,997	100%	922,181	100%	12,880	100%	144,869	100%	1,228,791	100%

Source: ES-202 Employment and Wages Program; 40 to 60 percent of all farmworkers are reported by employers subject to UI cov
Most states have Federal Unemployment Tax Act (FUTA) coverage of farm employers; farm employers who pay quarterly
    cash wages of $20,000 or more or employ 10 or more workers for at least one day in each of 20 different weeks must
    cover their farmworkers (20-10 rule). CA, ME, MN, RI, TX, and WA have more inclusive farmworker coverage.
Employers are farms or their subunits which employ 50 or more workers; this column is the number of ers reporting each quarter d
Ave annual employment is the employer-reported count of all workers on the payroll for the pay period which includes
    the 12th day of the month summed over 12 months and divided by 12; this column is the average of four quarterly reports.

218

Table A1.19
UI-Covered Employment and Wages Covered in Fruits and Vegetables: 1985(2)

State	Hort Specialt. Employers	Per Dist	Ave Annual Employmen	Per Dist	Annual Wages($000)	Per Dist	FLC Employers	Per Dist	Ave Annual Employment	Per Dist	Annual Wages($000)	Per Dist
Alabama	71	1%	1,943	1%	21,900	1%	6		211		2,158	
Alaska	5		89		1,104		0		0		0	
Arizona	39	1%	670	1%	9,180	1%	98	4%	4,858	5%	28,874	4%
Arkansas	22		198		1,507		10		16		247	
California	1,430	24%	31,374	24%	431,372	28%	1,039	37%	57,452	62%	438,601	64%
Colorado	88	1%	1,934	1%	23,068	1%	18	1%	308		1,838	
Connecticut	64	1%	2,200	2%	29,551	2%	0		0		0	
Delaware	13		254		2,195		7		68		880	
Florida	689	11%	18,379	14%	211,859	14%	886	32%	22,002	24%	158,133	23%
Georgia	76	1%	2,250	2%	21,620	1%	20	1%	193		2,989	
Hawaii	58	1%	947	1%	9,962	1%	0		0		0	
Idaho	29		484		6,002		4		74		191	
Illinois	231	4%	8,448	6%	77,405	5%	49	2%	213		4,554	1%
Indiana	176	3%	5,570	4%	51,727	3%	74	3%	1,174	1%	3,448	1%
Iowa	57	1%	670	1%	7,728		87	3%	249		4,556	1%
Kansas	45	1%	719	1%	8,286	1%	15	1%	19		428	
Kentucky	63	1%	637		6,247		0		0		0	
Lousiana	52	1%	727	1%	7,205		5		18		233	
Maine	8		70		702		2		5		60	
Maryland	46	1%	1,090	1%	12,594	1%	1		248		1,610	
Massachuse	78	1%	1,382	1%	18,564	1%	7		109		3,208	
Michigan	181	3%	3,509	3%	41,420	3%	7		153		834	
Minnesota	178	3%	1,684	1%	19,305	1%	13		54		729	
Mississippi	24		274		2,605		9		19		450	
Missouri	103	2%	1,734	1%	18,126	1%	18	1%	70		1,433	
Montana	13		198		1,780		0		0		0	
Nebraska	21		174		1,717		12		46		1,004	
Nevada	7		24		359		0		0		0	
New Hamps	25		333		4,764		0		0		0	

State	Hort Specialty Employers	Per Dist	Ave Annual Employment	Per Dist	Annual Wages($000)	Per Dist	FLC Employers	Per Dist	Ave Annual Employment	Per Dist	Annual Wages($000)	Per Dist
New Jersey	195	3%	2,517	2%	34,804	2%	6		54		433	
New Mexico	18		320		3,497		33	1%	690	1%	3,097	1%
New York	262	4%	3,289	3%	43,392	3%	9		112		1,334	
North Caroli	114	2%	1,835	1%	19,803	1%	29	1%	205	1%	1,213	
North Dakot	7		19		143		6		14		185	
Ohio	206	3%	4,815	4%	55,383	4%	8		23		253	
Oklahoma	45	1%	1,432	1%	16,826	1%	5		8		81	
Oregon	132	2%	4,099	3%	45,402	3%	15	1%	252	1%	1,538	
Pennsylvani	318	5%	8,571	7%	101,089	7%	6		55		547	
Puerto Rico	53	1%	539	1%	3,312		0		0		0	
Rhode Island	54	1%	525	1%	6,927		0		0		0	
South Carol	32	1%	911	1%	10,615	1%	10		105		592	
South Dako	9		85		713		8		35		700	
Tennessee	103	2%	2,100	2%	23,858	2%	4		14		344	
Texas	175	3%	5,299	4%	66,590	4%	223	8%	2,930	3%	12,663	2%
Utah	22		591		6,154		0		0		0	
Vermont	14		103		993		0		0		0	
Virginia	64	1%	1,169	1%	12,208	1%	4		16		129	
Virgin Island	0		0		0		0		0		0	
Washington	121	2%	3,169	2%	28,953	2%	11		164		719	
West Virgin	17		183		1,819		0		0		0	
Wisconsin	137	2%	1,418	1%	14,136	1%	9		108		3,436	1%
Wyoming	15		88		970		0		0		0	
United State	5,999	100%	131,036	100%	1,547,445	100%	2,775	100%	92,340	100%	683,723	100%

Source: ES-202 Employment and Wages Program; 40 to 60 percent of all farmworkers are reported by employers subject to UI coverage.

Most states have Federal Unemployment Tax Act (FUTA) coverage of farm employers; farm employers who pay quarterly cash wages of $20,000 or more or employ 10 or more workers for at least one day in each of 20 different weeks must cover their farmworkers (20-10 rule). CA, ME, MN, RI, TX, and WA have more inclusive farmworker coverage.

Employers are farms or their subunits which employ 50 or more workers; this column is the number of ers reporting each quarter divided by Ave annual employment is the employer-reported count of all workers on the payroll for the pay period which includes the 12th day of the month summed over 12 months and divided by 12; this column is the average of four quarterly reports.

Table A1.19

UI-Covered Employment and Wages Covered in Fruits and Vegetables: 1985(3)

State	FVH Employer	Per Dist	Ave Annual Employment	Per Dist	Annual Wages($000)	Per Dist	FVH Percent of all UI Crop Employment		
							Employers	Employment	Wages
Alabama	81		2,014	1%	22,348	1%	57%	73%	78%
Alaska	5	1%	89		1,104		70%	100%	100%
Arizona	144		6,082	2%	71,982	2%	27%	52%	51%
Arkansas	28		238		1,894		5%	6%	4%
California	11,737	54%	150,769	42%	1,631,637	44%	70%	76%	74%
Colorado	133	1%	2,760	1%	31,100	1%	52%	67%	67%
Connecticu	87		2,493	1%	32,988	1%	78%	55%	74%
Delaware	19		467		4,294		42%	73%	69%
Florida	1,652	8%	50,664	14%	504,721	14%	94%	92%	84%
Georgia	160	1%	4,011	1%	35,310	1%	53%	68%	66%
Hawaii	95		4,012	1%	54,180	1%	84%	49%	42%
Idaho	46		1,368		12,197		14%	23%	21%
Illinois	268	1%	9,285	3%	84,666	2%	51%	77%	69%
Indiana	210	1%	6,101	2%	56,274	2%	70%	90%	85%
Iowa	60		684		7,801		29%	34%	35%
Kansas	47		728		8,348		25%	46%	37%
Kentucky	63		637		6,247		47%	47%	47%
Lousiana	57		814		7,948		10%	16%	15%
Maine	44		496		5,778		52%	46%	54%
Maryland	64		1,401		15,371		45%	67%	69%
Massachuse	159	1%	2,637	1%	35,978	1%	88%	84%	90%
Michigan	382	2%	7,348	2%	72,994	2%	61%	71%	71%
Minnesota	216	1%	2,514	1%	26,172	1%	47%	61%	56%
Mississippi	27		327		3,168		5%	5%	6%
Missouri	140	1%	2,326	1%	21,760	1%	34%	64%	58%
Montana	18		205		1,847		20%	32%	25%
Nebraska	21		174		1,717		12%	11%	8%
Nevada	7		24		359		21%	5%	6%
New Hamp	33		614		6,955		96%	96%	96%

State	FVH Employers	Per Dist	Ave Annual Employment	Per Dist	Annual Wages($)	Per Dist	FVH Percent of all UI Crop Employment		
							Employers	Employment	Wages
New Jersey	622	3%	7,036	2%	78,982	2%	95%	98%	98%
New Mexico	57		1,368		14,481	3%	47%	56%	57%
New York	712	3%	9,214	3%	102,900	3%	88%	93%	93%
North Carolina	157	1%	2,550	1%	24,080	1%	37%	40%	45%
North Dakota	7		19		143		2%	2%	1%
Ohio	274	1%	6,431	2%	69,351	2%	62%	79%	77%
Oklahoma	51		1,502		17,537		55%	82%	80%
Oregon	422	2%	10,133	3%	85,774	2%	63%	74%	71%
Pennsylvania	429	2%	10,656	3%	118,894	3%	91%	96%	97%
Puerto Rico	1,081	5%	4,624	1%	13,388		44%	41%	43%
Rhode Island	77		603		7,513		82%	89%	90%
South Carolina	107		2,556	1%	21,451	1%	54%	66%	69%
South Dakota	9		85		713		28%	34%	24%
Tennessee	119	1%	2,298	1%	24,850	1%	62%	80%	82%
Texas	301	1%	12,249	3%	122,084	3%	41%	72%	68%
Utah	32		802		8,061		82%	92%	92%
Vermont	29		345		3,053		88%	92%	93%
Virginia	131	1%	2,841	1%	25,340	1%	66%	80%	78%
Virgin Islands	4		4		23		67%	55%	47%
Washington	887	4%	20,949	6%	143,919	4%	70%	78%	71%
West Virginia	46		847		6,549		96%	100%	100%
Wisconsin	273	1%	3,426	1%	41,225	1%	67%	59%	60%
Wyoming	15		88		970		61%	52%	54%
United States	21,838	100%	361,901	100%	3,698,417	100%	62%	72%	70%

Source: ES-202 Employment and Wages Program; 40 to 60 percent of all farmworkers are reported by employers subject to UI coverage.
Most states have Federal Unemployment Tax Act (FUTA) coverage of farm employers; farm employers who pay quarterly cash wages of $20,000 or more or employ 10 or more workers for at least one day in each of 20 different weeks must cover their farmworkers (20-10 rule). CA, ME, MN, RI, TX, and WA have more inclusive farmworker coverage.
Employers are farms or their subunits which employ 50 or more workers; this column is the number of ers reporting each quarter divided by fou
Ave annual employment is the employer-reported count of all workers on the payroll for the pay period which includes the 12th day of the month summed over 12 months and divided by 12; this column is the average of four quarterly reports.

Table A1.20
ES-223 Data: Migrant Farmworker Estimates for 1982

Migrant farmworkers do 25 to 149 days of farmwork and are unable to return to their homes at the end of the workday

State	Jan	Mar	Apr	May	June	July	Aug	Sept	Oct	Dec	Tot Worker Months	Per Dist	Ave Worker Months
Alabama				200	400	200	200	100	50		1,150		96
Arizona		1,900	1,700	1,600	2,200	900	1,100	1,300	1,700	1,900	14,300	2%	1,192
Arkansas													
California	16,600	15,300	21,900	22,800	26,600	30,600	29,200	36,000	29,100	12,600	240,700	32%	20,058
Colorado			100	300	1,600	2,000	2,300	1,800	800		8,900	1%	742
Connecticut													
Delaware				100	200	1,000	1,000	900	400		3,600		300
Florida	19,600	21,100	18,000	25,200	11,000	2,300	2,200	4,000	5,700	14,800	123,900	17%	10,325
Georgia			50	1,100	1,400	1,000	400	100	300		4,350	1%	363
Hawaii													
Idaho		300	300	1,000	2,400	3,500	3,900	3,200	3,500		17,800	2%	1,483
Illinois				400	800	300	700	600	400		3,200		267
Indiana				100	300	600	1,400	2,100			4,500	1%	375
Iowa				100	50	100	300	50	50		650		54
Kansas													
Kentucky													
Louisiana													
Maine					50	50	50		50		200		17
Maryland						1,300	1,200	400	100		3,000		250
Massachusetts				100	100	100	100	100	100		600		50
Michigan			600	2,500	6,800	11,200	14,900	12,000	10,200		58,200	8%	4,850
Minnesota					5,200	6,200	1,500				12,900	2%	1,075
Mississippi													
Missouri								300	100		400		33
Montana	200										200		17
Nebraska					900	700	300	100	100		2,100		175
Nevada													
New Hampshire								100	100		200		17

Migrant farmworkers are employed 25 to 140 days and are unable to return to their permanent homes

State	Jan	Mar	Apr	May	June	July	Aug	Sept	Oct	Dec	Tot Worker Months	Per Dist	Ave Worker Months
New Jersey	50		700	1,400	2,100	3,500	3,400	3,000	2,400	50	16,500	2%	1,375
New Mexico		50	50	300	3,000	2,300	1,100	200	400		7,500	1%	625
New York			700	700	1,400	2,600	3,500	5,300	6,700		20,200	3%	1,683
North Carolina	600	900	1,400	3,900	12,500	18,200	18,400	16,200	11,600	2,200	85,900	11%	7,158
North Dakota			100	400	1,500	1,300	400	400	400		4,500	1%	375
Ohio				400	1,100						1,500		125
Oklahoma	50	50	50	50	300	900	100	50	50	50	1,650		138
Oregon			200	1,000	2,800	2,800	1,900	3,900	1,300	50	13,950	2%	1,163
Pennsylvania	50		50	100	200	300	1,200	2,100	2,200	50	6,250	1%	521
Rhode Island													
South Carolina			200	1,300	4,000	1,700	500	300	300		8,300	1%	692
South Dakota													
Tennessee				500		700	700	200	100		2,200		183
Texas	300	400	200	1,200		3,100	2,600	400	400	300	9,100	1%	758
Utah				200	200	1,100	1,300	700	1,000		4,300	1%	358
Vermont								50	50		100		8
Virginia			200		400	2,400	2,500	1,700	1,400		8,600	1%	717
Washington	500	3,100	5,700			13,800	7,600	13,300	12,300	800	57,100	8%	4,758
West Virginia							300	600	600		1,500		125
Wisconsin													
Wyoming													
United States	37,950	42,800	51,300	65,450	91,200	116,750	106,250	111,550	93,950	32,800	750,000	100%	62,500

Source: ES-223 in-season Farm Labor reports, Employment and Training Administration, U.S. Department of Labor

This data is reported by job service offices and is based on a variety of data sources; reports of less than 50 workers are 50 in this table

Data are collected for multi-county agricultural reporting areas that employ 500 or more MSFW's or any H-2's

The farmworker estimate is the number of persons employed during the week which includes the 15th of the month;

a worker employed on two farms during the survey week could be counted twice

Seasonal workers are employed 25 to 149 days in farmwork and derive at least half of their total income from farmwork

Migrant workers are the subset of seasonal workers who are unable to return home at the end of the workday

Table A1.21

ES-223 Data: Seasonal Farmworker Estimates for 1982

Seasonal farmworkers are employed 25 to 140 days and obtain over half of their earned income from farmwork

State	Jan	Mar	Apr	May	June	July	Aug	Sept	Oct	Dec	Tot Worker Months	Per Dist	Ave Worker Months
Alabama				5,400	5,900	6,300	4,500	4,100	4,500	800	31,500	1%	2,625
Arizona		8,100	6,900	7,700	8,400	3,800	5,200	6,600	9,200	10,600	66,500	2%	5,542
Arkansas													
California	99,100	76,900	99,100	99,400	139,300	122,100	112,500	112,500	116,200	65,400	1,042,500	26%	86,875
Colorado			1,000	1,300	3,000	3,800	4,100	3,300	2,100		18,600		1,550
Connecticut			600	800	900	2,000	2,100	1,300	1,200	400	10,500		875
Delaware	600	600		2,600	2,800	5,400	5,400	4,700	2,600		23,500	1%	1,958
Florida	65,500	59,300	56,800	101,600	68,200	18,800	19,100	22,700	30,400	62,900	505,300	13%	42,108
Georgia		7,700	9,600	15,500	16,700	17,200	11,300	8,900	8,500		95,400	2%	7,950
Hawaii													
Idaho			1,300	3,000	5,100	6,400	7,400	6,100	8,600		37,900	1%	3,158
Illinois				1,000	1,800	4,100	2,100	1,800	1,900		12,700		1,058
Indiana				200	700	16,100	2,400	2,500			21,900	1%	1,825
Iowa				500	1,000	8,700	12,300	1,900	1,800		26,200	1%	2,183
Kansas													
Kentucky				3,400	5,600	5,800	5,500	9,000	7,200		36,500	1%	3,042
Louisiana													
Maine				700	1,200	1,100	4,600	2,300	4,900		14,800		1,233
Maryland						1,500	1,400	700	500		4,100		342
Massachusetts				1,600	1,800	2,100	1,900	1,700	1,300		10,400		867
Michigan			4,000	7,800	14,000	20,000	26,800	22,400	19,400		114,400	3%	9,533
Minnesota					5,200	6,200	1,500				12,900		1,075
Mississippi													
Missouri						800	200	900	200		2,100		175
Montana	1,800										1,800		150
Nebraska					2,000	2,100	5,300	1,200	900		11,500		958
Nevada													
New Hampshire								800	900		1,700		142

Seasonal farmworkers are employed 25 to 140 days and obtain over half of their earned income from farmwork

State	Jan	Mar	Apr	May	June	July	Aug	Sept	Oct	Dec	Tot Worker Months	Per Dist	Ave Worker Months
New Jersey			2,300	3,500	4,600	10,800	7,400	6,000	4,600		39,200	1%	3,267
New Mexico	400	500	500	700	4,000	3,200	2,100	1,200	1,200	600	14,400		1,200
New York				3,400	5,000	7,200	8,300	12,200	14,000		50,100	1%	4,175
North Carolina	53,800	53,900	59,800	69,900	80,400	102,700	110,900	101,100	84,000	59,700	776,200	19%	64,683
North Dakota			1,300	1,900	2,900	3,000	2,000	1,900	1,900		14,900		1,242
Ohio				6,200	7,300						13,500		1,125
Oklahoma	1,200	1,900	2,100	3,100	4,000	7,100	3,600	2,900	2,800	2,300	31,000	1%	2,583
Oregon	1,300	2,600	2,600	3,900	13,800	15,000	9,300	10,300	5,300	1,200	62,700	2%	5,225
Pennslyvania	4,500	4,000	4,800	6,300	6,500	7,300	7,800	8,700	8,100	3,200	61,200	2%	5,100
Rhode Island									50		50		4
South Carolina			600	2,400	8,400	5,000	2,200	1,500	1,300		21,400	1%	1,783
South Dakota													
Tennessee					21,300	22,700	21,800	30,600	17,000		113,400	3%	9,450
Texas	30,000	43,100	55,600	62,400	69,700	62,100	53,200	50,500	47,800	36,900	511,300	13%	42,608
Utah					500	1,900	2,600	1,500	1,600		8,100		675
Vermont								700	600		1,300		108
Virginia				1,100	2,200	6,200	6,900	6,500	4,400		27,300	1%	2,275
Washington	4,700	11,600	16,200			51,900	28,800	33,800	34,000	4,400	185,400	5%	15,450
West Virginia							500	1,200	1,800				
Wisconsin													
Wyoming													
United States	262,900	267,600	325,100	417,300	514,200	560,400	503,000	486,000	452,750	248,400	4,034,150	100%	336,179

Source: ES-223 in-season Farm Labor reports, Employment and Training Administration, U.S. Department of Labor

This data is reported by job service offices and is based on a variety of data sources; reports of less than 50 workers are 50 in this table

Data are collected for multi-county agricultural reporting areas that employ 500 or more MSFW's or any H-2's

The farmworker estimate is the number of persons employed during the week which includes the 15th of the month; a worker employed on two farms during the survey week could be counted twice

Seasonal workers are employed 25 to 149 days in farmwork and derive at least half of their total income from farmwork

Migrant workers are the subset of seasonal workers who are unable to return home at the end of the workday

Table A1.22
Migrant Education Enrollment and Funding: 1982

State	Migrant Students 5 to 17	Percent Interstate	FTE Students 5 to 17	Percent Interstate	Former Mig Students 5 to 17	Former Mig Student FTE 5 to 17	Migrant Educa Funds $000-	Per Dist
Alabama	2,155	44%	1,258	36%	1,249	933	1,003	
Alaska	185	46%	121	44%	23	12	314	3%
Arizona	7,983	72%	5,432	68%	7,229	6,653	6,561	2%
Arkansas	10,608	82%	3,957	64%	4,915	4,471	4,432	25%
California	75,523	53%	61,532	52%	54,038	49,059	62,281	1%
Colorado	2,824	86%	1,558	77%	1,917	1,789	2,601	1%
Connecticut	984	87%	609	86%	2,787	2,499	2,361	
Delaware	393	76%	251	70%	905	797	855	
Florida	26,467	83%	19,057	80%	15,872	14,959	17,907	7%
Georgia	3,853	50%	2,416	43%	4,060	3,554	3,095	1%
Idaho	5,016	77%	3,074	72%	3,126	2,840	3,035	1%
Illinois	2,104	81%	1,239	75%	1,314	1,063	1,751	1%
Indiana	2,105	96%	764	92%	674	661	873	
Iowa	193	93%	107	90%	230	190	207	
Kansas	2,260	67%	1,356	70%	712	782	1,089	1%
Kentucky	3,153	40%	1,955	37%	5,764	5,022	3,667	
Lousiana	2,898	50%	1,703	46%	6,553	6,532	6,111	2%
Maine	1,448	43%	967	41%	4,446	4,063	3,077	1%
Maryland	607	95%	258	92%	188	167	795	
Massachusetts	1,194	89%	871	89%	5,158	4,902	5,441	2%
Michigan	9,225	88%	4,558	80%	2,296	2,039	5,477	2%
Minnesota	3,215	99%	1,257	98%	190	150	1,672	1%
Mississippi	2,209	36%	1,530	33%	2,843	2,629	2,694	1%
Missouri	1,544	56%	1,021	48%	1,650	1,463	1,642	1%
Montana	1,052	100%	428	100%	30	44	730	
Nebraska	780	100%	396	100%	398	289	374	
Nevada	887	82%	583	82%	88	94	370	
New Hampshire	25	28%	22	29%			32	

States ranked by mig students 5 to 17 in 1982

State	%
California	32.4%
Texas	25.3%
Florida	10.0%
Washington	3.4%
Arizona	2.9%
Michigan	2.4%
Arkansas	2.1%
Oregon	1.8%
Idaho	1.6%
North Carolii	1.6%
Georgia	1.3%
Oklahoma	1.1%
Kentucky	1.0%
New York	1.0%
Lousiana	0.9%
Colorado	0.8%
Mississippi	0.8%
Ohio	0.8%
Kansas	0.7%
Alabama	0.7%
Minnesota	0.7%
Illinois	0.7%
New Mexico	0.6%
Missouri	0.5%
Maine	0.5%
Massachuset	0.5%
Indiana	0.4%
New Jersey	0.4%

State	Migrant Students 5 to 17	Percent Interstate	FTE Students 5 to 17	Percent Interstate	Former Mig Student 5 to 17	Former FTE Mig Students 5 to 17	Migrant Educa Funds $000-	Per Dist
New Jersey	980	58%	722	51%	2,746	2,491	2,681	1%
New Mexico	1,742	74%	1,055	71%	3,066	2,645	2,340	1%
New York	2,873	59%	1,851	47%	3,410	3,111	3,734	2%
North Carolina	5,460	61%	3,029	57%	9,083	6,794	8,336	3%
North Dakota	1,326	100%	554	100%	5	5	638	
Ohio	3,861	99%	1,428	97%	494	516	1,455	1%
Oklahoma	3,048	54%	2,100	55%	1,371	1,229	2,085	1%
Oregon	5,127	71%	3,413	64%	3,978	3,557	5,488	2%
Pennsylvania	1,003	81%	534	74%	2,363	2,094	2,143	1%
Rhode Island	22	86%	14	84%	43	46	14	
South Carolina	1,447	80%	540	70%	18	17	539	
South Dakota	73	100%	55	100%			31	
Tennessee	373	69%	180	60%	334	287	278	
Texas	66,231	58%	48,047	53%	74,302	68,111	63,133	26%
Utah	416	87%	268	83%	424	377	379	
Vermont	224	24%	198	19%	290	300	253	
Virginia	762	99%	329	99%	116	88	651	
Washington	9,627	68%	6,507	61%	6,666	6,094	9,350	4%
West Virginia	173	92%	83	90%	206	182	175	
Wisconsin	1,522	92%	679	88%	749	668	1,213	
Wyoming	632	97%	263	95%	47	45	292	
United States	277,812	65%	190,129	59%	238,366	216,312	245,653	100%

State	Per Dist
Wisconsin	0.4%
Connecticut	0.3%
Nevada	0.3%
North Dakota	0.3%
South Caroli	0.3%
Pennsylvania	0.3%
Montana	0.2%
Nebraska	0.2%
Virginia	0.2%
Utah	0.1%
Wyoming	0.1%
Maryland	0.1%
Delaware	0.1%
Vermont	0.1%
Tennessee	0.1%
Alaska	0.1%
Iowa	0.1%
West Virgini	0.1%
South Dakota	
New Hampsh	
Rhode Island	

Source: Office of Education, MSRTS Enrollment Summary

U. S. total excludes Puerto Rico, D.C., and other federal territories; there is no Migrant Education program in Hawaii

FTE is days of migrant student enrollment divided by 365

Migrant children had parents who moved across school district lines for temporary or seasonal agricultural or fishery jobs

Former migrant students had parents who were migrants sometime within the past 5 years

Table A1.23

Migrant Estimates Made in 1973 and 1986

State	Migrant Target Population Estimates for 1973		Lillisand Migrant Population Estimates for 1976			Rural America estimates for 1976		Where Have all the Farmworkers Gone?		
	Mig Pop	Per Dist	Mig Pop	Source	Per Dist	Mig Work	Per Dist	Mig Dep	Mig Pop	Per Dist
Alabama	1,890		4,813	DOL		1,290	1%	2,895	4,185	1%
Arizona	4,613	1%	17,714	MH	1%	3,300	1%	8,053	11,353	2%
Arkansas	5,274	1%	6,066	MH		125		179	304	
California	83,233	13%	244,949	DOL	16%	50,954	22%	105,542	156,496	24%
Colorado	11,392	2%	30,742	DOL	2%	4,631	2%	7,061	11,692	2%
Connecticut	5,179	1%	6,031	MH		1,863	1%	2,473	4,336	1%
Delaware	5,437	1%	9,379	MH	1%	1,354	1%	1,940	3,294	1%
Florida	76,450	12%	166,964	MH	11%	15,044	6%	30,254	45,298	7%
Georgia			31,558	DOL	2%	8,535	4%	18,901	27,436	4%
Hawaii				DOL						
Idaho	14,462	2%	25,134	MH	2%	3,446	1%	8,374	11,820	2%
Illinois	24,247	4%	41,826	MH	3%	2,122	1%	2,526	4,648	1%
Indiana	7,617	1%	20,449	DOL	1%	9,194	4%	11,601	20,795	3%
Iowa	1,411		2,435	MH		480		633	1,113	
Kansas	4,593	1%	8,924	MH	1%	5,430	2%	6,309	11,739	2%
Kentucky	186		618	DOL		250		368	618	
Lousiana	8,984	1%	10,332	MH	1%	63		113	176	
Maine	113		16,311	OE	1%	104		157	261	
Maryland	4,563	1%	7,871	MH	1%	1,320	1%	1,904	3,224	
Massachusetts	2,884		3,677	MH		971		1,350	2,321	
Michigan	51,776	8%	77,664	MH	5%	10,355	4%	13,280	23,635	4%
Minnesota	25,193	4%	43,457	MH	3%	7,115	3%	10,260	17,375	3%
Mississippi			20,078	OE	1%	3,500	1%	7,727	11,227	2%
Missouri	1,187		2,048	MH		306		357	663	
Montana	4,067	1%	17,250	DOL	1%	6,839	3%	10,016	16,855	3%
Nebraska	3,234		5,579	MH		1,172		1,334	2,506	
Nevada			616	OE		481		562	1,043	
New Hampshire	109		524	DOL		58		68	126	
New Jersey	11,146	2%	19,227	MH	1%	4,542	2%	6,410	10,952	2%
New Mexico	6,519	1%	7,715	DOL	1%	2,244	1%	4,465	6,709	1%

State	Migrant Target Population Estimates for 1973 Mig Pop	Per Dist	Lillisand Migrant Population Estimates for 1976 Mig Pop	Source	Per Dist	Where Have all the Farmworkers Gone? Rural America estimates for 1976 Mig Work	Per Dist	Mig Dep	Mig Pop	Per Dist
New York	13,380	2%	32,200	FSP	2%	5,942	3%	8,509	14,451	2%
North Carolina	6,101	1%	40,250	FSP	3%	13,841	6%	30,304	44,145	7%
North Dakota	5,719	1%	14,194	OE	1%	5,822	2%	9,172	14,994	2%
Ohio	19,433	3%	48,806	DOL	3%	14,215	6%	19,440	33,655	5%
Oklahoma	7,853	1%	13,550	MH	1%	1,748	1%	2,082	3,830	1%
Oregon	16,749	3%	41,431	FSP	3%	4,711	2%	5,737	10,448	2%
Pennsylvania	4,025	1%	8,714	DOL	1%	2,298	1%	3,262	5,560	1%
Rhode Island	158		171	MH						
South Carolina	6,585	1%	21,545	DOL	1%	4,079	2%	9,786	13,865	2%
South Dakota			185	DOL		500		764	1,264	
Tennessee			1,435	DOL		506		742	1,248	
Texas	153,731	24%	318,225	DOL	21%	7,454	3%	19,433	26,887	4%
Utah	4,377	1%	7,076	MH		1,627	1%	2,890	4,517	1%
Vermont			433	DOL		60		85	145	
Virginia	4,429	1%	12,455	DOL	1%	2,546	1%	5,208	7,754	1%
Washington	28,309	4%	70,743	DOL	5%	15,884	7%	21,201	37,085	6%
West Virginia	707		1,679	DOL		550		752	1,302	
Wisconsin	10,817	2%	19,166	DOL	1%	4,290	2%	6,037	10,327	2%
Wyoming	4,900	1%	9,132	DOL	1%	2,402	1%	3,353	5,755	1%
United States	653,032	100%	1,511,341		100%	235,563	100%	413,869	649,432	100%

Sources: Migrant Health Program Target Population Estimates, mimeo, 1973
Lillisand, D. et. al. An Estimate of the Number of Migrant and Seasonal
Farmworkers in the United States and Puerto Rico, 1977
Lillisand source is the source of the bottom-up data, e.g. DOL is ES-223 data,
MH are the 1973 Mig Health estimates, and FSP means farmworker service programs
Migrant Health and Lillisand also estimated migrants in Puerto Rico
Rural America, Where Have all the Farmworkers Gone?, 1976

Table A1.24
The Distribution of Migrant Worker Activity: 1982

State	COA Crop Labor Expend 1982-$000	Hourly Field Wage "July-82"	Expendits Div by Wages Hours or Activ	Dist of Expendits Per Dist	Less than 150 day Crop Worker	Per Dist	CPS-HFWF Migrant Workers	Per Dist	Final Dist
Alabama	36,363	$3.28	11,086	1%	19,036	1%	9,342	4%	2%
Arizona	140,090	$4.11	34,085	2%	28,139	1%	3,184	1%	2%
Arkansas	121,334	$3.58	33,892	2%	23,307	1%	2,231	1%	2%
California	1,681,436	$4.69	358,515	26%	607,601	22%	39,529	17%	23%
Colorado	48,556	$3.22	15,080	1%	21,837	1%	3,771	2%	1%
Connecticut	10,056	$3.55	2,833		7,211		276		
Delaware	8,084	$3.55	2,277		3,136		325		
Florida	472,408	$3.90	121,130	9%	112,293	4%	42,664	19%	10%
Georgia	90,000	$2.99	30,100	2%	49,795	2%	17,887	8%	4%
Hawaii	128,196	$5.88	21,802	2%	6,537		1,997	1%	1%
Idaho	100,754	$3.45	29,204	2%	44,532	2%	5,007	2%	2%
Illinois	134,405	$3.53	38,075	3%	85,081	3%	4,890	2%	3%
Indiana	69,711	$3.82	18,249	1%	46,661	2%	2,234	1%	1%
Iowa	91,027	$3.45	26,385	2%	77,510	3%	4,241	2%	2%
Kansas	72,429	$4.30	16,844	1%	32,510	1%	2,082	1%	1%
Kentucky	97,667	$3.09	31,607	2%	189,009	7%	3,496	2%	3%
Lousiana	82,883	$3.56	23,282	2%	18,809	1%	1,627	1%	1%
Maine	25,068	$3.55	7,061	1%	24,221	1%	362		1%
Maryland	23,583	$3.55	6,643		20,143	1%	766		1%
Massachusetts	16,708	$3.55	4,706		8,746		537		
Michigan	99,052	$3.82	25,930	2%	85,743	3%	2,839	1%	2%
Minnesota	88,420	$3.43	25,778	2%	62,983	2%	5,447	2%	2%
Mississippi	100,190	$3.18	31,506	2%	21,842	1%	1,557	1%	2%
Missouri	64,475	$4.18	15,425	1%	28,731	1%	1,520	1%	1%
Montana	40,777	$3.14	12,986	1%	12,027	1%	3,392	1%	1%
Nebraska	69,122	$3.86	17,907	1%	35,480	1%	2,088	1%	1%
Nevada	7,468	$3.14	2,378		1,469		310		1%

State	COA Crop Labor Expend 1982-000	Hourly Field Wage "July-82"	Expendits Div by Wages Hours or Activ	Dist of Expendits Per Dist	Less than 150 day Crop Worker	Per Dist	CPS-HFWF Migrant Workers	Per Dist	Final Dist
New Hampshire	4,252	$3.55	1,198		2,304		75		1%
New Jersey	45,104	$3.55	12,705	1%	17,762	1%	852		1%
New Mexico	29,509	$3.14	9,398	1%	9,456		1,341	1%	2%
New York	90,809	$3.56	25,508	2%	48,511	2%	1,676	1%	4%
North Carolina	183,623	$3.40	54,007	4%	207,800	8%	5,387	2%	1%
North Dakota	63,230	$3.87	16,339	1%	23,272	1%	1,965	1%	1%
Ohio	64,907	$4.02	16,146	1%	51,333	2%	2,115	1%	3%
Oklahoma	36,320	$3.87	9,385	1%	20,963	1%	3,823	2%	1%
Oregon	108,212	$3.66	29,566	2%	101,025	4%	4,657	2%	2%
Pennsylvania	44,645	$3.81	11,718	1%	32,328	1%	1,957	1%	5%
Rhode Island	992	$3.55	279		673		105		
South Carolina	53,336	$3.28	16,261	1%	40,324	1%	7,107	3%	2%
South Dakota	21,804	$3.87	5,634		12,332		866		2%
Tennessee	54,912	$3.28	16,741	1%	85,870	3%	2,455	1%	5%
Texas	274,717	$3.93	69,903	5%	84,227	3%	15,177	7%	
Utah	8,306	$3.14	2,645		9,710		534		
Vermont	2,280	$3.55	642		2,693		70		1%
Virginia	58,303	$3.25	17,939	1%	52,529	2%	1,549	1%	5%
Washington	262,910	$4.34	60,578	4%	205,624	8%	8,750	4%	
West Virginia	9,946	$3.55	2,802		5,707		113		1%
Wisconsin	61,758	$2.94	21,006	2%	37,069	1%	1,713	1%	
Wyoming	8,645	$3.14	2,753		3,487		461		
United States	5,409,270	$3.80	1,367,923	100%	2,730,046	100%	226,416	100%	100%

COA crop labor expenditures are the labor expenditures reported by crop farms in the 1982 COA

Hourly field wage is the average wage for the state in Farm Labor for July 1982; some of this data was revised in May 1985

Less than 150 day crop workers is the number of "seasonal" workers crop farmers reported hiring in the1982 COA

CPS-HFWF is the number of migrants in each region in the 1983 HFWF report distributed to states on the

basis of each states's crop sales in 1983

Index

Adverse Effect Wage Rate (AEWR), 34-35
AEWR. See Adverse Effect Wage Rate
Agricultural employment, UI-covered, 212-221(tables)
Agricultural Reporting Areas (ARAs), 62
Agricultural service firms, 12-13, 22, 38-39, 124
Agricultural service workers, 38
Agriculture, labor-intensive, 128-131
Aliens, illegal, 8, 131-132, 135. See also Special Agricultural Workers
ARAs. See Agricultural Reporting Areas

BEA. See Bureau of Economic Analysis
Bracero program, 6, 7
Bruce Church, 129
Bureau of Economic Analysis (BEA), 45-46, 47(table), 205-207(tables)
Bureau of Labor Statistics, 51

California, 4, 5, 12, 14, 15, 17, 31, 62, 71, 94, 100-101, 103-104, 108, 109, 110, 128, 132, 134-135, 138
CAMP. See College Assistance Migrant Program

Campbells, 129
Cash labor expenses, 44, 202-204(tables)
Caste of Despair, A (Goldfarb), 3, 111
Castle and Cooke, 128
Census of Agriculture (COA), 11, 15-16, 21, 26-34, 73(n2), 87, 180-199(tables)
Census of Population (COP), 2, 46, 48-49
1980, 50(table), 208-211(tables)
Centaur Associates, 94, 96-97, 112
Chavez, Cesar, 7, 126
COA. See Census of Agriculture
College Assistance Migrant Program (CAMP), 68, 162
Commission on Migrant Education, 70
Community facilities Loan Program, 156
Community Services Block Grant Program, 168
COP. See Census of Population
CPS. See Current Population Survey
Crop losses, 39
Current Population Survey (CPS), 49, 51-52, 53(table), 54, 106, 114, 115(table)

Data
  adjustments of, 74, 83, 90-91, 97, 104
  administrative, 54, 57, 98, 100
  client, 57
  ES-223, 222-225
  farm employment, 27-28
  farm operator, 27
  household, 46, 71
  household and dependent factors, 79, 81(table), 92-93, 118
  labor expenditure, 16, 28-29
  peak migrant estimates, 79-80
  QALS, 99
  sources of, 24, 25(fig.), 26, 57, 70, 88
  state-by-state, 179-231
  UI, 99-100
  UI, California, 102(table), 107, 116-117, 120-124, 133, 135
Democratic party, 6
Department of Agriculture (USDA), 11, 13
Department of Labor, 61
Depression, 5
Dole, 128, 129

Economic Indicators of the Farm Sector: State Financial Summary, 44
Economic Research Service (ERS), 42, 44, 202-203(table)
Elementary and Secondary Education Act, 65, 68, 70
Employment and Earnings, 51

Employment and Training Administration (ETA), 61, 62
Employment and Wages, 59
ERS. See Economic Research Service
Establishment data, 26
ES-223, 61-63, 64(table), 72(table), 87, 89-90, 106, 222-225(tables)
ETA. See Employment and training Administration
Even Start, 70

Fair Labor Standards Act, 38
Farm(s)
  cash grains, 194-195
  crop, 31, 198-199(table)
  definitions of, 9, 11, 21, 22
  fruit and nut, 124, 129, 188-189(table)
  fruit and vegetable, 14
  fruits, vegetables, and horticultural specialties (FVH), 31, 33(table), 34, 128, 129, 130, 192-193(table)
  horticulture specialty, 190-191(table)
  operators of, 9, 11, 23
  other field (013), 196-197(table)
  product sales of, 181-182(table)
  vegetable, 124, 129, 186-187
Farm Costs and Returns survey (FCRS), 26, 42, 204(table)
Farmers, 49, 50(table), 208-209(table)

Farm labor, 34, 36, 37, 40,
    41(table), 42, 43(table),
    200-201(table)
Farm labor distribution
    indicators, 72(table)
Farm Labor Housing Loan and
    Grant Program, 157
Farm Production Expenditures
    report, 44
Farmworkers
    advocates vs. employers of,
        127
    characteristics of, 139
    definitions of, 23, 59, 100
    distribution of, 72(table),
        73, 96, 97
    earnings of, 134
    employers of, 11, 127
    vs. farm jobs, 24
    groups of, 61
    harvest, 4-5
    hired, 11, 23, 31,
        32(table), 36-37,
        43(table), 53(table), 96,
        97, 182-183(table), 184-
        185(table)
    households of, 23-24
    1980 Census of Population,
        210-211(table)
    numbers of, 49, 50(table),
        97, 127, 139
    seasonal, 34, 37, 61,
        72(table), 89-90, 134,
        139(n2), 224-225(table)
    and seasonality, 20, 66
    unpaid, 36
    year-round, 133
    See also Farmworkers,
        migrant
Farmworkers, migrant, 14-18
    age of, 117-118
    Black, 112, 114, 115(table)

career-relevant data of, 117-
    118
casual, 123
characteristics of, 110-125
data on, 52, 53(table), 54
definitions of, 2-3, 8, 9,
    10(fig.), 11-14, 18,
    19(n3), 62, 77, 88, 100,
    101, 102(table), 103, 110
dependents, 80, 119. See
    also Data, household and
    dependent factors
distance travelled as,
    56(table), 114, 118, 124
distribution of, 14-16, 78,
    99, 103-107, 105(table),
    108, 109, 120, 230-
    231(table)
distribution formula for,
    104, 105(table), 106-107
earnings of, 54, 111, 113,
    114, 116-117, 120, 121-
    122
employers of, 123-124
estimates of (1973/1986),
    228-229(table)
families, 116
federal assistance programs
    for, 156-178
groups of, 1
as heads of households,
    114, 116
health care of. See Migrant
    Health program
Hispanic, 112-113, 114,
    115(table)
and immigration reform,
    131-138, 139
intra-ag, 101, 103, 107
and legal services, 116
life expectancy of, 113

myths about, 8, 110, 124, 125
1982 estimates of, 222-223(table)
nonfarm work of, 117, 121, 122
numbers of, 3, 8, 63, 78, 99, 100-101, 103, 107, 108, 112, 114, 120
and poverty lines, 116, 125
seasonal, 77, 80, 99, 121, 122, 123. See also Farmworkers, seasonal
skilled, 2
total, 103
White, 112, 114, 115(table), 119
work locations, 55(table)
See also Farmworkers
Farmworkers Justice Fund, 175
FCRS. See Farm Costs and Returns Survey
Fieldworkers, 40, 41(table)
Florida, 14, 15, 31, 63, 109, 110, 111
Fringe benefits, 38, 45
Fruit and vegetable farms, 14
UI-covered employment and wages, 216-221(tables)
Fruits, vegetables, and horticultural specialties (FVH), 31, 33(table), 34, 128, 129, 130, 192-193(table). See also under Farm(s)
Fuller, Varden, 3, 110
FVH. See Fruits, vegetables, and horticultural specialties

Galarza, Ernesto, 6
Goldfarb, Ron, 3, 111
Grapes of Wrath, The (Steinbeck), 2

Handicapped Migratory Agricultural and Seasonal Farmworker Vocational and Rehabilitation Services Program, 163
Harvest of Shame, 7, 111
HCR report, 84-87, 97
HEP. See High School Equivalency Program
HFWF. See Hired Farm Working Force
High School Equivalency Program (HEP), 68, 164
Hired Farm Working Force (HFWF), 13, 22, 51-52, 54, 96-97, 101, 106, 114, 115(table), 119
Hired hands, 4
Homestead Act, 4
H-2A program, 135-136, 174

Immigrants, 5
Immigration and Naturalization Service (INS), 135
Immigration reform, 131-138, 139
Immigration Reform and Control Act (IRCA), 39, 131-132, 135
INS. See Immigration and Naturalization Service, 135
In-Season Farm Labor Reports. See ES-223
InterAmerica Research Associates, 80

Inter/Intrastate Coordination (Section 143), 165
Interstate Migrant Education Council/Education Commission of the States, 177
IRCA. See Immigration Reform and Control Act

Jefferson, Thomas, 128
Job-turnover-ratio, 75, 98(n1)

Kentucky, 16, 87

Labor contractors, 130
Labor expenditures, 30(table), 44, 45-46, 47(table), 104, 105(table), 106, 180-181(table)
Labor expenses, 205-207(table)
Labor markets, fruit vs. vegetable, 129
Labor-saving machinery, 130-131
La Follette Committee, 5-6
Legal Services Corporation (LSC), 88
Lettuce, 128
Lillisand study, 82(table), 88-94, 95(table), 98(n3), 111-114, 118-119
LSC. See Legal Services Corporation
ME. See Migrant Education program
Methodology, 84-85, 87, 96, 107
Mexican-Americans, 111, 118
Mexicans, 5, 132
Michigan, 63

Migrant and Seasonal Farmworker Program, 172
Migrant and Seasonal Farmworker Services (MSFW), 173
Migrant and Seasonal Worker Protection Act (MSPA) Program, 171
Migrant Education program (ME), 13, 17-18, 65-68, 69(table), 70, 72(table), 226-227(table)
Migrant Employment and Training programs, 13
Migrant Head Start, 169
Migrant Health program, 13, 77-83, 85, 170
    Target Population Estimates report (1973), 78-80, 81(table), 82(table), 91, 93, 95(table)
    Target Population Estimates report (1978), 80, 82
Migrant Legal Action Programs, 88, 176
Migrants
    children, 65, 66-68, 69(table), 70
    dependents, 75, 92
    distribution of, 14-16, 17
    diversity of, 8
    family, 9, 68
    farmworkers. See Farmworkers, migrant
    fieldworkers, 40
    Hispanic, 3, 93
    numbers of, 96-97
    semi-skilled/professional, 9
    students, 67

studies, 77-83
total, 63
Migrant Student Record
  Transfer System
  (MSTRS), 57, 65-68, 71
Migration patterns, 113
MSFW. See Migrant and
  Seasonal Farmworker
  Services
MSPA. See Migrant and
  Seasonal Worker
  Protection Act Program
MSRTS. See Migrant Student
  Record Transfer System

NASS. See National
  Agricultural Statistics
  Service
National Agricultural Statistics
  Service (NASS), 35, 44
National Governor's
  Association, 178
National Labor Relations Act, 6
National School Breakfast
  Program (NSBP), 158
National School Lunch
  Program (NSLP), 159
Nazario, Sonia, 111
North Carolina, 16, 63
NSBP. See National School
  Breakfast Program
NSLP. See National School
  Lunch Program

OSHA, See Occupational
  Safety and Health
  Administration

Pesticide Farm Safety Program,
  167
Public Health Service Act, 77
Puerto Rico, 179

QALS. See Quarterly
  Agricultural Labor Survey
Quarterly Agricultural Labor
  Survey (QALS), 26, 34-
  42

RAW. See Replenishment
  Agricultural Worker
Regions, standard federal, 120
Replenishment Agricultural
  Worker (RAW), 135-138
Republican party, 7
Rural America study, 88, 93-
  94, 95(table)

Salvage theory, 122-123
SAS. See Seasonal
  Agricultural Services
SAW. See Special Agricultural
  Workers
School Improvement Act
  (H.R.5), 68, 70
Seasonal Agricultural Services
  (SAS), 39, 132, 133-
  134, 136-138
Senate Committee on Education
  and Labor. See La
  Follette Committee
Sharecroppers, 11
SIC. See Standard Industrial
  Classification
SMP. See Special Milk
  Program
Social Security Numbers
  (SSNs), 100, 135
Special Agricultural Workers
  (SAW), 132-133, 134-
  135, 136-138
Special Milk Program (SMP),
  160

Special Supplemental Food
Program for Women,
Infants and Children
(WIC), 161
SSNs. See Social Security
Numbers
Standard Industrial
Classification (SIC), 21,
100
Standard Industry Codes, 11
State Basic Grant Program
(Section 141), 166
Steinbeck, John 2
Strawberries, 128-129
Sunkist, 128
Sweatshops in the Sun, 111

Temporary Alien Agricultural
Labor Certification
Program (H2-A), 135-
136, 174
Texas, 15, 63, 109, 110, 113
Tree-shakers, 130
20-10 rule, 16, 58

UI. See Unemployment
Insurance
Unemployment Insurance (UI),
11, 16-17, 58-59,
60(table), 61, 72(table),
212-221(tables). See also
Data, UI; Data, UI,
California
United Farm Workers, 7
USDA. See Department of
Agriculture
Wages
annual, 200-201
cash grains farm, 194-
195(table)

field and livestock workers,
200-201(table)
fruit and nut farm, 188-
189(table)
fruits and vegetables, 216-
221(tables)
fruit, vegetables, and
horticultural (FVH) farm,
192-193(table)
general crop farm, 198-
199(table)
horticulture specialty farm,
190-191(table)
hourly, 37-38, 40,
41(table), 105(table),
120, 121, 122, 134
1982, 205(table)
nonfarm, 121
other field (013), 196-
197(table)
UI-covered, 60(table), 212-
221(tables)
vegetable farm, 186-
187(table)
See also Farmworkers,
earnings of;
Farmworkers, migrant,
earnings of
Wall Street Journal, 111
Washington, 109, 138
Wheat, 4, 128-129
"Where Have All the
Farmworkers Gone?",93
WIC. See Special
Supplemental Food
Program for Women,
Infants and Children
World War II, 6